P9-BYQ-598

fP

PRIOR BOOKS BY KEVIN DAVIES

Cracking the Genome:
Inside the Race to Unlock Human DNA

Breakthrough:
The Race to Find the Breast Cancer Gene
(with Michael White)

The $1,000 GENOME

THE REVOLUTION IN DNA SEQUENCING AND THE NEW ERA OF PERSONALIZED MEDICINE

Kevin Davies

Free Press
NEW YORK LONDON TORONTO SYDNEY NEW DELHI

fP
FREE PRESS
An Imprint of Simon & Schuster, Inc.
1230 Avenue of the Americas
New York, NY 10020

Copyright © 2010 by Kevin Davies

All rights reserved, including the right to reproduce this book or portions thereof in any form whatsoever. For information address Free Press Subsidiary Rights Department, 1230 Avenue of the Americas, New York, NY 10020.

First Free Press trade paperback edition August 2015

FREE PRESS and colophon are trademarks of Simon & Schuster, Inc.

For information about special discounts for bulk purchases, please contact Simon & Schuster Special Sales at 1-866-506-1949 or business@simonandschuster.com.

The Simon & Schuster Speakers Bureau can bring authors to your live event. For more information or to book an event, contact the Simon & Schuster Speakers Bureau at 1-866-248-3049 or visit our website at www.simonspeakers.com.

Designed by Paul Dippolito

Manufactured in the United States of America

1 3 5 7 9 10 8 6 4 2

The Library of Congress has cataloged the hardcover edition as follows:
Davies, Kevin
The $1,000 genome: the revolution in DNA sequencing and the new era of personalized medicine / Kevin Davies.
Includes bibliographical references and index.
1. Human Genome Project. 2. Human gene mapping. I. Title.
II. Title: One thousand dollar genome.
QH445.2.D368 2010
616'.042—dc22 2010007317

ISBN 978-1-4165-6959-6
ISBN 978-1-4165-6961-9 (pbk)
ISBN 978-1-4165-7018-9 (ebook)

To Susan . . . and all my fellow
R1b1b2a1a2f haplogroup members

Contents

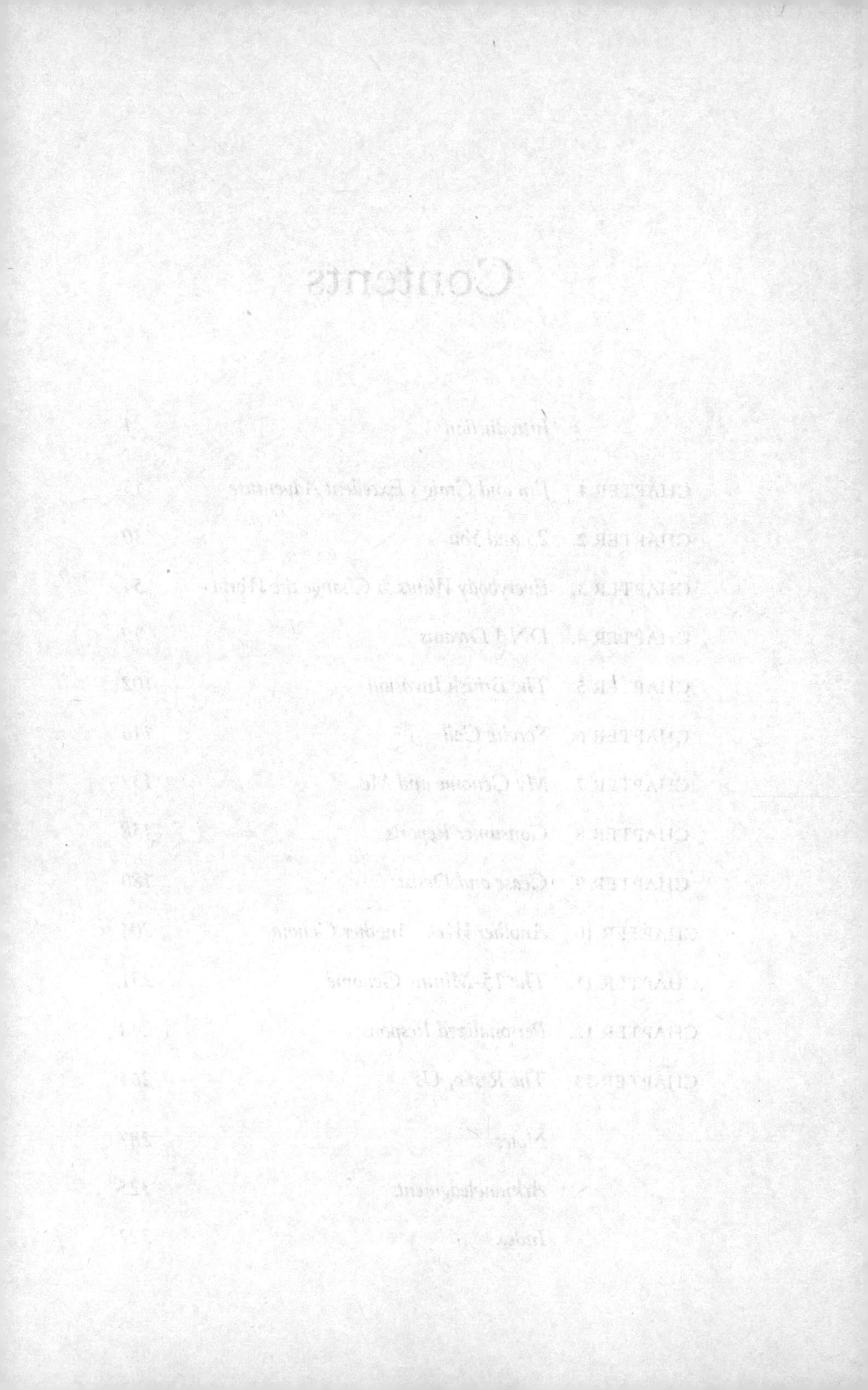

Introduction

Other than a terminal case of hair loss, Jeffrey Gulcher MD had little cause for concern when he swiped a sterile spatula inside his mouth for a peek into his DNA. Gulcher was the forty-eight-year-old cofounder and chief science officer of deCODE Genetics, an Icelandic biotech company intent on harnessing the extraordinary history of the island's tiny, isolated population to pinpoint disease-related genes and develop new drugs. He might be carrying a few extra pounds, and his boss could be a handful at times, but otherwise Gulcher was in good shape.

In November 2007, deCODE had launched a new Web-based, direct-to-consumer (DTC) service called deCODEme. It was the first DTC personal-genomics service on the market, and a test cost just under $1,000. deCODEme was not a genetic test for a specific disease nor a genealogical foray into a client's ancestry but a tiny sampling of the 3 billion letters (in the familiar four-letter alphabet, A, C, T, and G) of that person's genetic code. Some variations from this sequence—an A or a T out of place—correlated with increased risk of common diseases, including cancers, arthritis, diabetes, and heart disease. For less than the price of a flat-screen TV, anyone could read their hereditary horoscope. No prescription required.

Gulcher followed the same simple directions he'd demonstrated in a YouTube video.[1] From his cheek swab, deCODE purified his DNA and analyzed some 600,000 discrete letters in his genome. He was not surprised to learn he had a genetic predisposition to baldness and average risks for a dozen other medical conditions. But a handful of those DNA markers suggested that Gulcher had double the average lifetime risk for type 2 diabetes and prostate cancer. He addressed the former by purchasing a treadmill. The latter, however, was less straightforward. Gulcher's genotype (his genetic makeup) predicted a one-in-three lifetime risk of prostate cancer—and an aggressive form of the cancer at that. Gulcher's father had been diagnosed at the age of seventy with prostate cancer, but that was hardly unusual. Gulcher's prostate-specific antigen (PSA) screen was 2.5, normal, if a little high, for a man under fifty. His doctor referred Gulcher to famed Northwestern University urologist William Catalona, the pioneer of PSA testing and a renowned prostate surgeon. Catalona had even collaborated on some of deCODE's groundbreaking prostate cancer genetics research.

Catalona performed an ultrasound-guided biopsy, which revealed that Gulcher had a grade 6 (Gleason scale) prostate carcinoma. "I'm not only the CSO [Chief Scientific Officer], I'm also a client," Gulcher joked the week after his biopsy. "This test may have saved his life," deCODE CEO Kari Stefansson told me in all seriousness shortly before the surgery to remove Gulcher's tumor. The tumor had been caught early, and there were no signs of metastasis into the bone. Had it not been for the deCODEme gene scan, Gulcher would not have checked his PSA, his doctor would not have referred him to a specialist, and the urologist might not have performed a biopsy. And while his surgery and attendant care cost about $20,000, Gulcher reckoned his preventative measures saved the American taxpayer about $140,000 in long-term medical care.

Around the same time that Gulcher was swabbing his cheek, another personal-genomics start-up called Navigenics was hosting a launch party in the unlikely surroundings of the SoHo district in New York City. TV talk-show host Maury Povich was in attendance, supporting his son-in-law, David Agus, a Navigenics cofounder and a prominent Los Angeles oncologist. His coworkers affectionately dubbed Agus "Doc Hollywood" for his celebrity clientele. Al Gore, who had flown in just for the reception, called him "a miracle worker." Agus admitted that he had received his own wake-up call by simply spitting into a tube and, like Gulcher, having his DNA analyzed. The results revealed that Agus had an 80 percent lifetime risk of a heart attack, twice the national average. He was now taking appropriate lifestyle and medical measures, and his daughters had banned him from eating French fries. He also mentioned the story of a young woman who had volunteered for a colonoscopy after receiving her Navigenics results, which revealed a heightened risk for colon cancer. The doctors discovered a three-centimeter polyp, which was removed.

By far the most intriguing and publicized of the consumer-genomics start-ups was Silicon Valley's 23andMe, named for the 23 pairs of chromosomes (the bundles of DNA found in the nucleus of our cells) that carry and transmit our genetic material. Neither of the company's founders, Anne Wojcicki and Linda Avey, was a trained geneticist or doctor or had run a biotech company, but Wojcicki was married to Google cofounder Sergey Brin, which counted for something. 23andMe collected celebrity saliva at fashionable "spit parties," racked up magazine spreads and network TV appearances, and Skyped into *Oprah*. *Time* magazine voted the company's retail DNA test the Invention of the Year for 2008.[2] When Brin tried the test, he considered it a largely hypothetical exercise. But his major health concern was Parkinson's disease (PD), which affected his mother and a great-aunt. Parkinson's is not typically inherited in the straightforward dominant or recessive (Mendelian) manner of muscular dystrophy or Huntington's disease, but mutations in a few genes are associated with rare hereditary cases of PD, particularly in families of Ashkenazi Jewish descent, and Brin is of this lineage. The list of traits and conditions evaluated by 23andMe did not initially include PD. But Brin, like any customer, could search his raw DNA data for variations in just about any of the roughly 20,000 genes in his genome. When he typed in *LRRK2*, a gene that can have mutations associated with PD, he found he carried a known mutation called G2109S, which he later confirmed his mother had as well.[3] Brin's lifetime risk of contracting Parkinson's is uncertain, but is put at about 60 percent. He publicized his results in a blog, but far from being upset or worried, Brin felt he was in a unique and fortunate position:

> I know early in my life something I am substantially predisposed to. I now have the opportunity to adjust my life to reduce those odds (e.g. there is evidence that exercise may be protective against Parkinson's). I also have the opportunity to perform and support research into this disease long before it may affect me. And, regardless of my own health it can help my family members as well as others. I feel fortunate to be in this position. Until the fountain of youth is discovered, all of us will have some conditions in our old age only we don't know what they will be. I have a better guess than almost anyone else for what ills may be mine—and I have decades to prepare for it.[4]

It would be easy to dismiss the awakenings of Gulcher, Agus, and Brin as the predictable self-serving pronouncements of scientific, medical, and Internet entrepreneurs with inextricable personal and financial ties to the birth of the personal-genomics industry. For many other curious personal-genome clients and pioneers, including myself, the news is anything but black and white. Unlike a traditional genetic test, the new tests deal in probabilities, odds ratios, and relative risks for common diseases that afflict us all and that include genetic and environmental components, neither of which is well defined. It's like a TV weather reporter predicting a 60 percent chance of rain, when all one really wants to know is: Do I need my umbrella tomorrow or not?

But this takes nothing away from the efforts to provide personal genomic information to any consumer who wants it. For as little as $399, the personal-genomics pioneers—I'll refer to them collectively as 23 et al.[5]—offered the public an unprecedented opportunity to peek at our unique genetic code, learn about our ancestry, and fathom what disorders we might develop in the future. Some clients were attracted by the recreational aspects of the service: Can I smell asparagus in my urine? Am I a fast or slow caffeine metabolizer? Do I have a genetic predisposition to heroin addiction? Others were curious about tracing their family's roots or comparing genomes with friends and family members. Adoptees wanted to glean something of their biological background. And doubtless some were simply sold on the prospect of having something cool to discuss at cocktail parties.

The main attraction, however, was to discover useful information on health and disease susceptibility. While 23 et al. insisted that their services were not in any way medical diagnostic tests but were simply for educational or informational purposes, Brin's experience in particular highlighted the extraordinary potential of consumer genomics. Here was a billionaire talking about his medical prognosis based not on a physical exam or a brain scan but a solitary glitch he uncovered in his own DNA with a few taps on his computer keyboard. And if Brin could make life-changing decisions based on scanning less than 0.1 percent of his DNA, imagine what might happen when anyone could receive their entire genome on a DVD or flash drive.

By and large, the medical establishment reacted with disdain to this outpouring of genomic democratization. Doctors, ethicists, regulators, and medical journal editors rushed to condemn the wisdom, validity, and legality of delivering complex, unsubstantiated medical information to consumers, at least without the guiding hand of an expert medical practitioner. Some worried that a positive risk for an incurable disease would trigger panic, maybe even thoughts of suicide. Others feared the opposite—that a genetically clean bill of health would give consumers a false sense of security. It reminded me of the scene in the film *Groundhog Day,* when Bill Murray's character learns he's invincible:

Rita [Andie MacDowell]: Don't you worry about cholesterol, lung cancer, love handles?
Phil [Bill Murray]: I don't worry about anything anymore.
Rita: What makes you so special? Everybody worries about something.
Phil: That's exactly what makes me so special. I don't even have to floss!

Rather than fork out $500 or $1,000 for a DNA scan, the editors of the *New England Journal of Medicine* argued, it made more sense to buy a gym membership or change one's diet to reduce the risk of heart disease and diabetes.[6] The states of New York and California cracked down on consumer-genomics services. Boston University neurologist and genetic epidemiologist Robert Green worried about a nightmare scenario in which a woman ordered a personal-genome scan that was unable to distinguish the *BRCA1* test for hereditary breast cancer from a group of other cancer gene markers that influence risk in the general population. "Oh my God, I've got the gene for breast cancer," she might think, while her surgeon might misinterpret the results: "Yeah, you're positive for the gene, I guess we better schedule surgery."[7]

The advocates for personal genomics made a couple of powerful arguments in rebuttal. One concerned personal education and empowerment. Who should stand between you and the right to see information about your own genetic code? The Food and Drug Administration or some meddling state government bureaucrat? Your doctor or insurance company? Many consumers are better-informed about modern genetics than their family physician. In his classic book *Genome*, published shortly before the completion of the Human Genome Project, the British author Matt Ridley warned of the specter of employers, insurers, medical, and government bureaucrats meddling in people's genetic affairs. He wrote:

I am keen not to share my genetic code with my insurer, I am keen that my doctors should know it and use it, but I am

> adamant to the point of fanaticism that it is my decision. My genome is my property and not the state's. It is not for the government to decide with whom I may share the contents of my genes. It is not for the government to decide whether I may have the test done. It is for me. There is a terrible, paternalistic tendency to think that "we" must have one policy on this matter, and that government must lay down rules about how much of your own genetic code you may see and whom you may show it to. It is yours, not the government's, and you should always remember that.[8]

The argument is that the right to see your DNA sequence is a matter of equity, of justice, and of basic civil rights.

Another powerful argument in favor of DNA testing, espoused by Navigenics cofounder Dietrich Stephan, is the abysmal state of medical economics in the developed world and the imminent health care crisis. Four million Americans have Alzheimer's disease, costing the health care system a staggering $500 billion a year. By 2050, the number of patients will be approaching 20 million, the annual cost a ridiculous $1.5 trillion to 2 trillion for a single disease. "Suddenly, you're paying forty percent GDP [gross domestic product] for health care. That is simply unsustainable," said Stephan.[9] In 2005, the United States spent $2 trillion on health care, a figure that is projected to reach $4 trillion—or 20 percent of GDP—by 2015. By 2020, Medicare could be gone. By 2035, there will be a national health care crisis that can only be averted through the practice of prevention. But according to the World Health Organization, despite topping the charts in health care expenditure, the United States barely scrapes into the top forty nations in terms of quality. Take the case of type 2 diabetes. The United States spends $132 billion in managing this type of diabetes annually. A diabetic patient will spend about five times more on medical care than a nondiabetic, but with early diagnosis, even modest lifestyle improvements could greatly reduce the incidence and cost of treating the disease and its complications.[10]

The arrival of personal-genome analysis and DNA sequencing is also a critical component for the emergence of personalized medicine. Knowing a patient's genetic makeup in advance could allow a doctor to identify whether that patient would respond poorly or adversely to a drug, helping prevent a repeat of the Vioxx scandal, where a blockbuster drug had to be withdrawn because of a few fatal side effects. It could prevent people suffering fatal gastrointestinal bleeding from taking aspirin or hemorrhaging to death because they are prescribed the wrong dose of a blood thinner. They could avoid the trauma of electroshock treatment, which may be resorted to because doctors are unable to prescribe the right dose of an antidepressant, let alone select the most appropriate drug.

The purveyors of personal genomics insist that genomic medicine will—indeed, *must*—become an integral part of modern health care. "I think it is very likely that, within [a few] years, pretty much every college-educated person in the United States is going to have a profile similar to the one provided by deCODEme or 23andMe or Navigenics," said Kari Stefansson.[11]

Dietrich Stephan of Navigenics believes personal genomics will become part and parcel of twenty-first-century medicine—a complete personal-genome sequence, a mini Human Genome Project for every person, quite possibly performed at birth. "Ultimately, every baby that's born will and should have their genome sequenced," Stephan said. Doctors will routinely "sequence the genome, put it in a big computer, push a button, and get a rank-ordered list of things you are at risk for. It will supplant newborn screening and all molecular diagnostics."[12]

In Stephan's vision of twenty-first-century medicine, a baby's genome is sequenced, and then predispositions are unmasked at appropriate stages with parental approval, starting with a modified version of the newborn screening panel, which is currently used to test for dozens of diseases and inborn errors. Later, at eighteen or twenty-one, in collaboration with his or her physician, Stephan envisions unmasking all the other risk predictions and then implementing "a lifestyle and behavior-modification plan, a biomarker monitoring program, and an early detection paradigm. Rather than just go for your physical, stick this thermometer in your mouth, take your blood pressure, there's a whole personalized aspect to that physical. That's the goal." In the near future, the price for that full personalized genome sequence will tumble as low as $1,000. "Things will start to explode from there," he said.

While the arrival of 23 et al. provided the first tantalizing taste of personal genomics, something was missing: 99.9 percent of the total DNA sequence, or thereabouts. What if people could be offered the sequence of their entire genome, all 3 billion letters of DNA, just as two legendary American scientists, Craig Venter and James Watson, received in 2007? Shortly after the launch of 23andMe, a Boston start-up called Knome launched the first genome-sequencing service for the rest of us. Knome (like "gnome") offered millionaire clients their genome on a fancy USB stick for a cool $350,000. But with the price of genome sequencing in free fall thanks to the rampant spread of next-generation sequencing technologies, it became inevitable that the $1,000 genome—the geneticists' holy grail—would be reached sooner than anyone expected.

The term "$1,000 genome" was coined in 2001 and captured the imagination of scientists in 2002, when Craig Venter hosted a symposium to showcase new technologies poised to bring it about. And who better? "Darth Venter" had been the media's bad boy in the race for the Human Genome Project (HGP), which I described in my previous book, *Cracking the Genome.*[13] The original Washington, DC, maverick, at least in scientific circles, Venter had mounted a hostile takeover of the HGP as CEO of Celera Genomics, which sought to become the "Bloomberg of genomics," referring to the popular financial news and data service. He hoped to sell multimillion-dollar DNA database licenses to pharmaceutical companies. President Clinton finally helped arrange a tepid truce between Venter and HGP consortium leader Francis Collins in June 2000, allowing Venter and the HGP consortium to share credit for helping decipher what Clinton called "the language in which God created life." The final cost of the HGP—from the 1990 kickoff to formal completion on the fiftieth anniversary of the discovery of the double helix—was a staggering $2.7 billion. The actual sequencing effort in the final two years probably cost $500 million on both sides.

Even before the HGP celebrations, it was clear that the classic first-generation DNA sequencing technology, named after British Nobel Prize winner Fred Sanger, was just about tapped out. It had taken factories of one hundred or more machines many months just to sequence one human genome. Something faster and cheaper was desperately needed. From New England to Olde England, scientists, engineers, and entrepreneurs were drawing blueprints on pub napkins, finding inspiration in hospital waiting rooms, and writing patents and business plans in their basements and dorm rooms. The race to build the next generation of sequencing machines—one that could match the throughput of an entire academic or biotech genome center in a single, small machine and slash the cost of a genome from millions of dollars to mere thousands—was under way.

The star attraction at Venter's $1,000 genome 2002 symposium was a superconfident young Asian-American named Eugene Chan, a medical school dropout who audaciously promised that his company, U.S. Genomics, would be able to sequence a human genome in less than 60 minutes. The era of affordable personal genomics suddenly seemed within reach—and the HGP wasn't even done yet. The "$1,000 genome" instantly became a catchphrase that perfectly captured the ultimate goal of next-generation DNA sequencing and the medical revolution that would unfurl.

The first such platform was launched commercially by 454 Life Sciences in 2005 and delivered the personal-genome sequence of Jim Watson in 2007. At $1 million, this was a remarkable improvement on the billions spent on the HGP. By 2009, a quartet of next-gen manufacturers were vying for market dominance, with a dozen or more pretenders waiting to emerge. "Here's how we think of the $1,000 genome," said Stanley Lapidus, founder and chairman of Helicos Biosciences. "Anything between $100 and $10,000 is the $1,000 genome. Maybe even $20,000 or $25,000 is the $1,000 genome." In other words, it's not literally $1,000 that matters, but slashing the cost of a person's genome to the point where it is closer to the price of an MRI than a Maserati.

In the summer of 2009, Illumina, the leader in the sequencing market, launched a new service to sequence anyone's genome for just $48,000 and a doctor's prescription. CEO Jay Flatley wasn't expecting a flood of orders, but he was eager to lay the groundwork for a future where personal-genome sequencing was commonplace, if not routine. At the end of 2009, Complete Genomics sequenced the genome of a Harvard professor for just $1,500, while another emerging company, Pacific Biosciences, promised to deliver the "15-minute genome" by 2014.

This astonishing triumph of technology is still a little hard to take for someone who remembers how painful and tedious DNA sequencing used to be. Somewhere in the subterranean tunnels of the off-site depository of Senate House, the University of London library, sits a dusty copy of my PhD thesis. This literary masterpiece failed to make a lasting, or even fleeting, contribution to the annals of medical science, but it serves as a nostalgic artifact of a golden age in human genetics in the 1980s, just prior to the launch of the HGP. Scientists tasted the first success in dropping molecular pins on rudimentory maps of the twenty-three pairs of human chromosomes to help guide the search for the rogue genes underlying inherited disorders such as Huntington's disease, muscular dystrophy, and cystic fibrosis (CF). Finding one mutant gene among the roughly 25,000 genes strewn across the human genome was like searching for the proverbial needle in a haystack. In case some people didn't get the metaphor, Francis Collins even posed for a newspaper photographer in his lab coat on top of a real haystack at a farm, clutching a large needle.

Our group at London's St. Mary's Hospital was involved in an international race to map and isolate the gene for CF, the most common genetic disease among Europeans. From time to time, our group leader, a ginger-mopped Lancastrian named Pete Scambler, would gather the troops in the local pub for a pint and a pep talk. "There's a rumor the CF gene is on chromosome 13," he said one lunchtime. I took a dejected swig of bitter, fearing I'd wasted a year rummaging through the wrong end of the haystack.[14]

I eventually managed to sequence a respectable thousand bases of DNA at a cost of about $1 per base. At that rate, it would have taken a million years to sequence the human genome (explaining my hasty retreat from the bench a short time later).

A lot can happen in twenty-five years, of course. In the 1980s, many PhDs were awarded for piecing together the sequence of a single gene, just a few thousand bases. The first human genome took thirteen years to assemble. By 2009, that task took less than fourteen days and just $1,500. Now in 2010, we are on the brink of decoding someone's genome for a mere $1,000. "Within five years," Venter said in his contribution to the BBC's televised series of annual Dimbleby lectures in 2007, "it will be commonplace to have your own genome sequenced—something that just a decade ago required billions of pounds and was considered a monumental achievement." The prospect of routine affordable genome sequencing is inextricably linked with the future of health care—a vision of medicine that, as Lee Hood, the grandfather of automated DNA sequencing, says, is personalized, predictive, preventative, and participatory. Sensibly, he shortened it to "P4 medicine."

In *The $1,000 Genome*, I tell the story of the people and technologies that are transforming the practice of medicine by providing the means to have your own personal genome sequenced. We meet the people behind the birth of personal genomics, the application of these services, what they can tell consumers about their health and disease liabilities—and just as important, what they cannot. These organizations have democratized genetics by bringing valuable information directly to the consumer while raising a host of medical, ethical, and legal questions. I share insights from my own genome results and critique the various consumer-genomics services.

I also examine the birth of next-generation sequencing, the technology that will be the centerpiece of biomedical research and preventative medicine. We meet the visionaries who are devising methods for sequencing DNA at astonishing speeds previously imagined only by science fiction writers. Thanks to their efforts and ingenuity, the cost of sequencing a human genome has plummeted from billions of dollars for the HGP to a couple of thousand dollars in 2010. That's a remarkable drop of a million-fold, or a factor of 10 every year, far outstripping Moore's law (Intel cofounder Gordon Moore's 1965 maxim that the density of transistors on a computer chip doubles every 18–24 months). And that's before a new wave of third-generation sequencing technologies emerge—machines and possibly even portable battery-powered Tricorder-like devices that can sequence a human genome in fifteen minutes for $100.

This technology will drive the mass sequencing of the population in the very near future, possibly at birth. I also tackle the question of what this personal-genomics revolution means for you. We explore what information can be learned, separate fact from fiction, and look at how you can use this information in your personal life, from decisions about health and lifestyle to drug selection.

The $1,000 genome is a watershed moment in medicine and human history; we are the species that has learned to unravel our genetic code and are now on the verge of expanding that achievement to millions of people, and potentially almost everyone on the planet. The value of that information is limited right now, but there can be no doubt that simultaneous advances in countless branches of science from mathematics to computational biology to neuroscience will help make sense of our genetic instruction manual in ways we can barely fathom today.

On June 26, 2000, President Clinton officially marked the completion of the first draft of the human genome, an encyclopedic reference for medical science and a historic achievement for mankind. Remarkably, in just a decade, the sequencing of individuals with cancer and other diseases, from all corners of the globe, has become so affordable as to be an almost weekly occurrence. As I intend to show in *The $1,000 Genome*, we are on the precipice of being able to routinely sequence and interpret just about any human genome, including, one day, yours. Let the journey begin.

CHAPTER 1

Jim and Craig's Excellent Adventure

On the morning of May 31, 2007, eight scientists and physicians squeezed shoulder to shoulder across a narrow makeshift stage at the Baylor College of Medicine in Houston in readiness for a historic press conference.[1] Despite the significance of the occasion, they almost outnumbered the smattering of journalists in attendance. The guest of honor needed no introduction: the seventy-nine-year-old doyen of DNA himself, James Watson, the man who codiscovered the iconic double helix, wrote the classic book of the same name, and shared the Nobel Prize in 1962. By comparison, the other newsmaker was unknown: a wealthy, boyish biotech entrepreneur named Jonathan Rothberg. Two years earlier, Rothberg had nervously invited Watson to have his complete DNA sequenced using a radical new lightning-speed technology developed by his company, 454 Life Sciences.

When Watson, Rothberg, and other members of the "Project Jim" team reassembled later that afternoon, the room was packed with scientists, physicians, and students. After ninety minutes of speeches, there was a short, poignant ceremony. "I have one last very pleasurable duty," announced Baylor Genome Center director

Richard Gibbs in his melodic Australian accent. It was Gibbs who had urged Rothberg to approach Watson and who had conducted the detailed analysis of Watson's sequence. Just as data had flown "from Watson's blood cells to 454 to Baylor," Gibbs said it was only appropriate now to reverse the flow. He handed a small package wrapped in a red ribbon to Rothberg, who in turn presented it to Watson. It was a portable hard drive—there was just too much raw data to burn it onto a DVD—containing a digital text of some 24 billion characters in the trademark four-letter alphabet of DNA, A, C, G, T. Even the most famous biologist of the twentieth century would have difficulty deciphering its meaning. It was, said Gibbs, simply "the first personal genome."

Well, almost. Just days earlier, J. Craig Venter, Watson's nemesis throughout the heyday of the Human Genome Project (HGP), had electronically transmitted his own genome sequence to Genbank, the official DNA sequence archive of the National Institutes of Health (NIH). Watson didn't seem to care unduly about precedence. Regardless of whose DNA sequence was technically deposited into the database first, Watson was the first person on the planet to reap the rewards of next-generation sequencing—a revolution in speed-reading DNA that promised to do for genome analysis and personalized medicine what the microprocessor did for computing. The price tag was falling incredibly fast. The HGP took thirteen years and cost $2.7 billion. The preliminary analysis of Watson's genome took thirteen weeks and a mere $1 million. Fittingly, the man who launched the genetics revolution in 1953 with Francis Crick by piecing together the pairs of bases that constitute the rungs of the double helix was present at the birth of the personal-genomics revolution half a century later.

For Rothberg, presenting Watson, one of his heroes, with his digital genome was the culmination of a dream that had begun eight years earlier—an intensely personal quest to build a technology that would usher in a new world of personalized medicine. "My work on the personal genome started with . . . my three beautiful children," Rothberg said that afternoon. When his first child was diagnosed with tuberous sclerosis, Rothberg responded by building a nonprofit institute for childhood diseases. In the summer of 1999, his second child, Noah, was a "blue baby" because of deoxygenation of the blood and was rushed to the newborn intensive care unit. "I was upset," said Rothberg. "I was mostly upset because we didn't know what was wrong. I was upset for a lack of information. . . . Why can't I have complete information on Noah? Why can't I have Noah's genome? Because if I had his genome, we'd know, with the physicians, what to worry about and what not to worry about."

That night, as Rothberg pondered the hundreds of millions of dollars being lavished on the HGP with little to show for it at the time, it occurred to him that somebody needed to do for DNA sequencing what Jack Kilby and Robert Noyce had done for the computer industry in the 1950s by inventing the integrated circuit.[2] "If we could do what the computer industry had done [with] personal computers, we could in fact sequence individual genomes. That night, my vision was to create a chip to sequence individual and personal genomes." Rothberg had the vision for a company that he cryptically called 454 Life Sciences.

The first proof that 454's technology could work came in 2003, when it successfully sequenced the DNA of a tiny virus. Rothberg flashed a slide depicting two contrasting quotes from the ensuing *New York Times* story covering that event, knowing it would amuse the Baylor audience.[3] "This is going to be big," Gibbs, the local favorite, had judged. But another noted geneticist, the U.S. Department of Energy's Eddy Rubin, said skeptically, "I think doing a whole bacterium will be a challenge." Two summers later, 454 proved Rubin wrong by doing just that. "An innovation is taking an invention and making it practical for others to use," Rothberg said. From there, 454 had embarked on the Neanderthal genome. "A Neanderthal never crossed a body of water if he did not see land on the other side," Rothberg continued. "But as soon as we had *Homo sapiens* coming out of Africa, we were on Easter Island. There was no adventure we didn't undertake." Laying the Neanderthal genome alongside its *sapiens* cousin would reveal "the genes involved in that adventure-seeking, that creativity." Then, with a neat piece of comic timing, Rothberg segued: "If you can sequence Neanderthal, it's quite possible you could sequence anybody," juxtaposing a picture of our extinct cousin with a photograph of a young Jim Watson.

Rothberg and Gibbs hatched Project Jim in early 2005 over dinner. "Let's sequence a human!" Rothberg urged his companions, half hoping they'd choose him. Gibbs suggested Watson instead and offered to give him a call. A week later, Rothberg was standing in line at a pharmacy in a Boston mall when his cell phone rang. "Hold for Jim Watson." After making arrangements to visit Watson, Rothberg hung up and immediately called his mother. "I just talked to Jim Watson!" he said proudly. Prior to traveling to Cold Spring Harbor, Rothberg rehearsed his pitch. Getting Watson's informed consent would be essential, but surely nobody understood DNA better than the man who codiscovered the double helix. Awestruck, as most visitors are upon entering Watson's office, with its framed awards and honorary degrees lining the walls, and the Nobel medal on the desk, Rothberg was shaking. But even before he could finish his spiel, Watson agreed. Rothberg had Watson at "Hello."

It took more than a year before 454 seriously embarked on sequencing Watson's DNA. 454 churned out a total of 24 billion As, Cs, Gs and Ts from Watson's DNA.[4] That meant that, on average, they had sampled each of the three billion letters in Watson's genome eight times for eightfold coverage. It was a massive genomic jigsaw puzzle with 106 million pieces; each piece, or fragment, contained an average of just 230 bases of DNA. Rothberg realized his team was in trouble barely one billion bases into the project. There's no record of him calling Gibbs to say, "Houston, we have a problem," but he admitted that the Baylor group came to his rescue. Gibbs assigned the informatics challenge—assembling the genomic jigsaw and poring over the long tail of DNA variants—to his colleague David Wheeler. Sequencing Watson's DNA was one thing; trying to make sense of it and give him some medically useful information was another. For one thing, surprisingly, 3.5 percent of

Watson's DNA sequence appeared to be novel, with no matches to the reference genome produced in the HGP.[5]

At the press conference Rothberg praised Watson's example in releasing his sequence publicly, which meant "we don't have to be afraid." Five years from now, Rothberg said, before having a child, couples would look at their partners' lists of genetic flaws to see which faulty genes overlap. "If you're an Ashkenazi Jew, you look for Tay-Sachs. In the future, you'll look at both genomes to mitigate that risk."

Rothberg ended his speech with a punch list of his major takeaways from Project Jim, which went like this: we all have significant genomic variation from each other, changes in single letters, sentences, and entire pages of the book of life. Watson's genome inventory, for example, revealed 310 genes with likely mutations and 23 with known disease-causing mutations, increasing his risk for cancer and heart disease. The Baylor team recommended that he should take folic acid and other vitamins and minimize his exposure to sunlight, particularly during his daily tennis matches.

James Watson, in his eightieth year, looked particularly fit and resplendent in a cream suit and blue checkered shirt as he shuffled to the microphone. Like Rothberg's eureka moment in creating 454, Watson said his motivation for launching the HGP was also personal. In 1985, Watson hosted a talk by the Nobel laureate Renato Dulbecco at Cold Spring Harbor Laboratory in which he proposed an all-out program to sequence the human genome as an essential strategy in the war against cancer. But by the time Dulbecco published his ideas in *Science*,[6] Watson was more interested in mental illness than cancer. In 1986, Watson's oldest son, Rufus, began displaying problems and was sent home from boarding school. Later, he ran away to Manhattan and traveled to the top of the World Trade Center, where he tried to break a window. He was hospitalized with suspected schizophrenia, and the lives of Watson and his wife, Elizabeth, were forever altered. "It seems to me it would be a hopeless task to ever understand schizophrenia or any other complex mental disorder unless we had the complete human genome sequence," said Watson.

Although proud of the HGP, Watson said it would not change medicine decisively until personal genomes became really cheap. "454 and other next-gen sequencing technologies are going to change the face of medicine extraordinarily," he predicted. "First, we really should have a DNA diagnosis of all cancers before we start treating them. . . . [The research] is going far too slow. If you're suffering from cancer, are these guys working on Sunday?"[7]

But Watson's biggest hope was to find the root cause of bipolar disorder and schizophrenia, citing progress in tracking genetic glitches in some patients with autism. He wanted Congress to earmark $1 billion to 2 billion to conquer mental illness: "Knowing what's wrong in mental disease won't cure my son, but it'll be a start, and you can always hope for that lucky break that he's suffering from a defective gene for which we actually know how to regulate its activities." He continued, "We're going to sequence DNA from mentally ill people until we find out what's wrong with them. No child should be getting medicine for bipolar disease when treated for mental disease if they don't have it."

Watson had more or less forgotten about Project Jim after donating a blood sample in 2005, and confessed he wasn't even sure initially if he'd be alive to enjoy the results. His initial intent was to have his entire DNA sequence posted on the Web, but he had a change of heart. There was one gene, the apolipoprotein E (*APOE*) gene on chromosome 19, that he would rather not know about.[8] About 2 percent of the population carry two copies of the E4 version of the gene, which imposes a fifteenfold increased risk of Alzheimer's disease. "Since we can't do much about Alzheimer's disease, I didn't want to know if I was at risk," said Watson. "My grandmother died of Alzheimer's disease at the age of 85. So I had a one-in-four chance of sharing the wrong form of that gene." The genotype of *APOE* was duly redacted from the public release of Watson's sequence.[9] There was a scare, however; a savvy bioinformatician, Michael Cariaso, along with others, inferred Watson's genotype from the pattern of markers in the neighboring gene, but did not disclose what he had learned.[10] (Cold Spring Harbor later scrubbed the sequence of the adjacent gene, *APOC*, as well.)

If Watson was hoping for instant enlightenment from his million-dollar sequence, he was disappointed. There were no genetic clues as to his susceptibility to basal cell carcinoma, which he had first contracted at the age of twenty-eight, or his son's illness. A major surprise was the news that he carried a defective version of *BRCA1*—the hereditary breast and ovarian cancer gene—which increases the risk of cancer in both sexes (albeit to a much lower degree in men than women).[11] "My sister had breast cancer at age 50," Watson divulged. "I take comfort in the fact that it largely affects women. I don't have a daughter—I never thought that was a good thing until today!" Watson smirked at the audience, oblivious to the bewilderment on people's faces.

He did have two nieces who had reason to be concerned. However, it turned out that the Baylor team had jumped the gun. Because of lingering uncertainty about the relevance of his putative *BRCA1* mutation, Watson later called the leading authority on breast cancer genetics, Mary-Claire King, the geneticist who first put *BRCA1* on the genetic map, in 1990. King had documented hundreds of mutations in *BRCA1* over the years, but Watson's wasn't one of them. She told Watson that what he had was likely nothing more than a benign Irish DNA variation.

The Baylor team informed Watson that he had twenty-three known disease-causing mutations in all. They had matched his unique DNA sequence with databases that catalogue thousands of known disease genes.[12] "People's genetic load is a lot more than most people recognize," said James Lupski, the clinical geneticist on the team. "Everyone carries between 10–20 recessive lethal genes, but you never know what you carry unless you marry someone who carries the same [mutated genes]."[13] Watson said the information about his genes would be difficult to interpret until scientists could line up thousands more genomes, "so we can correlate our physical and mental qualities with our genome." The greater significance was a symbolic one; it was the milestone that this "gigantic step" marked in the road to routine genomic diagnosis. Watson said, "I think we'll have a healthier and more compassionate world 50 years from now due to this technological advantage [that] we're celebrating."

Watson couldn't resist speculating on how individual genome data might be applied in the future. Here he waded into the minefield of eugenics, a field that Cold Spring Harbor Laboratory has a dubious history with. "Personal genome sequencing also brings enormous ethical challenges," he said. "[A] family won't let someone marry their daughter until they look at her prospective husband's genome. Will your genome complement my daughter's? It sounds bizarre, but I can imagine that happening in India or Israel. . . . It's too important to take chances with your children. It sounds like eugenics, [but] the aim is to have the next generation healthy."

Gibbs and his team had grappled with ethical considerations in releasing Watson's sequence: for example, whether he should have sought permission from his two children before going public with it. Amy McGuire, the Baylor bioethicist on the Project Jim team, urged Watson to consider the privacy implications and be respectful of his sons' wishes, but Watson hadn't seen any harm in plowing ahead. He wanted to set an example to demystify the process and to discourage people from hoarding their sequences "as if they're nuclear weapons."

McGuire, Lupski, and the Project Jim team leaders dined with Watson the night before the press conference to discuss his results. Sitting next to Watson, Lupski felt slightly uncomfortable as he was carrying the results of additional clinical array tests he had performed on Watson's DNA. They were performed under the stringent guidelines of the Health Insurance Portability and Accountability Act of 1997 (HIPAA), which established protections for personal health information. "This is patient information," Lupski said. "I'm not gonna divulge it to anybody, including my colleagues, until I've discussed it with my patient." Lupski suggested to Watson that they go somewhere private before presenting him with an envelope full of printouts and forms necessary to ensure HIPAA compliance. It's one thing discussing DNA mutations and gene numbers with a fellow geneticist, and quite another explaining the fundamentals of dominant and recessive genes to the general public. "The reality is, most individuals and patients are not Jim Watson," said Lupski. "How are you gonna explain these things to the public?"

Lupski's own expertise is in another layer of genomic variation called copy number variants (CNVs). There are at least 1,500 CNVs—segments of chromosomes that are duplicated or lost in different people scattered around the human genome. It's only since the HGP wrapped up that scientists have settled on a precise estimate of the total number of human genes[14]—about 20,500—and have appreciated the full extent of CNVs, which dramatically increase the amount of genetic variation between individuals and complicate the analysis of any personal genome.

Lupski gives the following illustration of a CNV. If a stretch of DNA reads:

I like to swim in the ocean but I do not like to swim in the pool.

Then some people might be missing a chunk:

I like to swim in the . . . pool.

Or they might have an extra copy:

I like to swim in the ocean but I do not like to swim in the ocean but I do not like to swim in the pool.

Most genetics textbooks, even those written after the HGP was completed, state that the amount of genetic variation between any two unrelated people is one base per thousand, or 0.1 percent. That is, we each possess variations in about 3 million bases out of the 3 billion in the human genome. These individual variations are called single nucleotide polymorphisms (SNPs) in the trade. But laid end to end, CNVs account for much more DNA, almost a full 1 percent of the genome. The significance of CNVs in human health is only just beginning to be appreciated, but there are tantalizing clues that CNVs contribute to autism, mental illness, and many other diseases.

The analysis of Watson's genome sequence was led by bioinformatics expert David Wheeler. Though he was an expert in analyzing DNA sequences, Wheeler had never been tasked with reading an individual's entire genome, let alone that of a living legend. By the fall of 2007, Wheeler's group had identified more than 9,000 "nonsynonymous" SNPs—DNA substitutions that can potentially affect gene function. Wheeler's team also found another 350 genes disrupted or impacted by small insertions or deletions of DNA (indels) and CNVs. But no problems seemed to occur in genes that were immediately medically important, or if they did, Watson was heterozygous, meaning he had one good copy of the gene.

Based on his groundbreaking adventure with Watson's genome, I expected Wheeler to be an unabashed advocate of personal genomics. But he refused to get carried away: "It's definitely premature from the perspective of providing the average person useful information," he said.[15] "It's going to take a while before we know how to synthesize all the information about the variation in a person. . . . Just sorting out what all the factors are that put somebody over the edge to get a disease is going to take a while." Wheeler felt that Watson was hoping to get an immediate genetic explanation for his predilections for skin cancer or political incorrectness. "I think he was disappointed," Wheeler told me later.

It took almost a full year before Watson's genome was finally published in *Nature,*[16] fittingly the same prestigious journal that published Watson and Francis Crick's seminal discovery of the double helix fifty-four years earlier. The respected University of Washington geneticist Maynard Olson summed up in this way the limitations of what the data revealed.

> If Watson took his sequence to a genetic counsellor [*sic*], there would be little to discuss. The sequence seems to show that he

> is a carrier for a handful of mutations that might catch a counsellor's interest. But these mutations have no known effects on Watson himself, and would confer risk on offspring only in the highly unlikely event of a marriage between two carriers. None of these mutations is ever likely to be considered an appropriate candidate for screening in the general population—of which, for these purposes, Watson is a representative member.[17]

Another interested party offered his own unique perspective on the Watson genome. "It's a new standard of sequencing technology, but I don't think it's a new standard of genome coverage," said one Craig Venter.[18] That's because Venter had already published his own genome sequence fully nine months earlier.[19]

As said before, Venter's goal in founding Celera Genomics in 1998 was to build "the Bloomberg of genomics." Celera would be the gatekeeper of the most advanced and sophisticated copy of the human genome and make a small fortune by granting licenses to Big Pharma companies. Celera's stock soared to $250 a share before collapsing in the 2000 crash. Meanwhile the forces behind the HGP consortium—aghast at the prospect of a genome monopoly—pushed ahead at an accelerated clip. On June 26, 2000, flanked by Francis Collins and Craig Venter, and with then–British Prime Minister Tony Blair cheerleading via satellite, President Clinton declared a tie of sorts. Eight months later, Venter and Collins published their draft genome reports simultaneously in *Science* and *Nature,* respectively, though there was still much work to do in finishing the sequencing.

Venter had made the decision to sequence his own genome almost a decade before Watson. He told me about it in a conversation in October 2002, the morning after hosting his symposium on the $1,000 genome. Earlier that year, he had quit Celera Genomics, and that very week his research institute had grabbed the covers of both *Nature* and *Science* after cracking the genomes of the malaria parasite and its host mosquito. As he tucked into a late breakfast of bagel and cream cheese, under the watchful eye of his publicist (and now wife), Heather Kowalski, I asked Venter why he'd revealed on *60 Minutes II* that his DNA was the primary source for the Celera genome sequence.[20] Venter said he'd initially hidden his identity because of the media circus surrounding the genome wars. "How many days was I on the front page of the newspapers for minor utterances? I didn't want there to be a circus and a field day around what, to me, was a very personal decision—a decision about trying to show leadership in the community—that you don't need to be afraid of this data." He continued:

> To me, there are two types of leadership. In Vietnam, there were the leaders that pushed the small guy out in front to be the point man to step on the booby traps and get shot first. And there are the leaders who actually led and got people to follow them. I've never been one to push people out in front of me to get shot first. I was a [DNA] donor out of just absolute scientific curiosity. My view is how can anybody possibly work in this field and not want to know their genetic code? Something's wrong with them! I mean, what the hell are they doing?! It's the ultimate dishonesty. They're advocating other people should do this. "We're going to interpret your life but, boy, stay away from mine?" I'm not shy about that. I wanted to know my own genetic code to understand my own life.[21]

Venter learned that he carried one *APOE4* marker for Alzheimer's. He wasn't worried about Alzheimer's, though, but rather heart disease, which is influenced by a person's *APOE* genotype. Venter's father had died at the age of fifty-nine from sudden cardiac arrest. "You don't have to be a genius to know you have some kind of genetic history there," he said. "It makes a big difference psychologically to know what you really have." He began taking a statin drug, which he'd been contemplating for a while.

After quitting Celera in early 2002, Venter returned to academia and established the eponymous J. Craig Venter Institute, where the

British geneticist Samuel Levy led the effort to complete sequencing Venter's genome; relying still on traditional Sanger sequencing. In October 2006, Venter offered the *Wall Street Journal* a sneak peek.[22] In addition to *APOE*, Venter had found DNA variants responsible for his penetrating blue eyes (*OCA2* on chromosome 15) and increased risk of macular degeneration (*CFH*). On the other hand, he did not carry variants associated with thrill-seeking (the dopamine D4 receptor) or melanoma (the p16 gene). Neither Venter's lifelong love of sailing and adventure nor an episode of skin cancer could easily be pinned on his genes.

Levy was in no great hurry to publish the Venter sequencing paper, but in March 2007, Venter's then deputy, Bob Strausberg, heard 454's Michael Egholm deliver a preview of Project Jim[23] and hurriedly called Levy, worried that they might get scooped in publishing the first individual genome. On May 9—six days before the Baylor group began uploading Watson's DNA files to the NIH—Levy and his colleagues submitted their paper to the journal *PLoS Biology*, and Venter's personal-genome sequence—dubbed "HuRef"—was published in September 2007.[24] Levy and friends hoped their "definitive molecular portrait" of Venter's genome would provide "a starting point for future genome comparisons" and introduce an era of individualized genomic information. Comparing Venter's sequence to the HGP reference, Levy's team catalogued 3.2 million SNPs, but that was only 25 percent of the DNA variation between the two genomes.

Venter's genome dramatically showed that we do *not* inherit two pristine copies of each chromosome from our parents. In many instances, entire genes are deleted (or duplicated) on one chromosome—the CNVs. For example, Venter lacks one copy of the *TSMG1* gene, which codes for a protein involved in metabolizing environmental toxins. Levy speculated there could be a connection between that missing gene and Venter's history of asthma and cancer, although he acknowledged, "It's hard to establish that cause and effect. We've only touched the tip of what's there."

While the data allowed Levy to rule out predisposing mutations to disorders such as Huntington's disease and CF that are caused by individual genes, interpreting mutations for complex diseases such as heart disease and diabetes and behaviors such as novelty-seeking was far trickier. Some variants suggested an increased risk; others seemed to mitigate that risk. The more genetic information that was available, the more complex the decoding of that information was becoming. An important next step was to compare the data from Venter and Watson, which was first done by Pauline Ng, a bubbly Venter Institute researcher. One area of interest was the sharply different ability of the two men to metabolize drugs.[25] Ng looked at *CYP2D6*, a gene that codes for an enzyme that metabolizes beta blockers, antidepressants, and other drugs. Watson had a variant called *10, associated with lower activity and found in fewer than 1 percent of Caucasians, whereas Venter had the more active *1 variant, technically known as an allele. This was a vivid example of the stark differences in drug response that can be elucidated by the genetic analysis of individuals, rather than lumping all members of a particular race or ethnic group into a single category on the fallacious assumption that everyone will respond the same. Watson used this information to lower his dose of beta blockers (to control blood pressure) from daily to weekly.[26]

In December 2007, Venter became only the third American to deliver the prestigious annual Richard Dimbleby Lecture, broadcast live by BBC.[27] If the twentieth century was the nuclear age, Venter proposed that the twenty-first century would be fundamentally shaped by advances in biology and genomics. Medicine and health care had to move toward a preventive philosophy, he said. "The only alternative to preventative medicine as the path to lowering health care costs would be to deny access altogether, which is clearly not acceptable." The prospects for preventive genomic medicine were undoubtedly exciting, but many questions loomed about how the field would develop.

The news that Venter and Watson were the first members of the personal-genome club gave some people pause over the issue of access to sequencing. "Is this the next space tourism?" one scientist demanded of Egholm shortly before the Houston press conference releasing Watson's data. "I'd hate the availability of single-genome sequencing to be based purely on money and fame," commented Cambridge University geneticist Michael Ashburner. "Just doing famous or very rich people is bloody tacky, actually."[28]

But back in Houston, Rothberg predicted that sequencing throughput would grow at least fivefold over the coming years. "The same way you saw your computers at WalMart for less than $1,000, you'll see the $1,000 genome," he said. The release of Venter's and Watson's data was the start of the inexorable, inevitable collapse in cost of individual genome sequencing, now progressing at an astonishing rate.

"Pioneers and explorers say 'yes,' " Rothberg said before handing Watson his digital DNA sequence. "This is the end of one quest—the dream to sequence a genome—and the beginning of another quest"—to develop and perfect the methods to routinely sequence an individual's genome for $1,000 or less, and to learn precisely what the sequence of those billions of base pairs means for one's health, behavior, and sense of self.

Just a few months after receiving his personal genome, Watson resigned as chancellor of Cold Spring Harbor Laboratory following publication of racially insensitive remarks in a British newspaper. "Watson is a brilliant scientist with a remarkable life story [but] along the way, he stepped on more than one landmine," observed Maynard Olson in *Nature*. Future historians could scrutinize Watson's sequence for clues to his character, but they would have to rely "on what Watson wrote, said and did during his lifetime rather than on the order of the base pairs in his genome."[29]

Perhaps the value of the first two personal genomes—both belonging to wealthy, white scientists—was indeed more symbolic than scientific. But a new movement was taking root committed to democratizing genomics and empowering consumers by providing them with personal genetic information, all for less than $1,000—and a small sample of spit.

CHAPTER 2

23 and You

"Genetics is about to get personal. Don't panic, we're here to help."

—23ANDME WEB SITE, NOVEMBER 2007

At the annual World Economic Forum, a surreal gathering of presidents and royalty, billionaires and Nobel laureates, rock stars and supermodels converge on the Swiss resort of Davos to discuss a litany of weighty global sociopolitical issues—when they aren't skiing and partying heavily. Among those attending the January 2008 event were the founders of Google, Larry Page and Sergey Brin, accompanied on the Google corporate jet by Brin's wife, Anne Wojcicki, her business partner Linda Avey, and, hitching a ride, Craig Venter.

Avey and Wojcicki weren't just tagging along to enjoy the après-ski. Rather, the cofounders of what they called the first personal-genome company were preparing the European launch of their start-up, 23andMe. In a cramped hallway at the posh Hotel Belvedere, the women, dressed in pristine white lab coats, coaxed the rich and famous—and a handful of journalists—to spit into sterile tubes. Few would have blinked at the $999 list price, but here the saliva kits were free. Among those summoning their saliva were supermodel Naomi Campbell, Dell Computers founder Michael Dell, Pulitzer Prize–winning author Thomas Friedman, and rock star Peter Gabriel.

Journalists usually prefer imbibing to expectorating, but they seldom turn down free swag. Among those visiting the 23andMe booth was *Financial Times* journalist Gideon Rachman, who blogged facetiously about what he might uncover about his health and ancestry. "Could I be Jewish? What if I am related to [fellow *FT* journalist] John Gapper? Am I genetically prone to obesity, or do I just eat too much?" He was particularly intrigued by the comely young lab technician who tutored him in the art of spitting. It was Wojcicki. "Presumably she has enough money never to have to handle bottled spit again in her lifetime," Rachman wrote. "It's impressive she keeps at it."[1]

The report from another British journalist, the *Daily Telegraph*'s Felix Lowe, appeared under the gloriously tabloid-worthy headline: "Google Wife Targets World DNA Domination."[2] Wojcicki was quoted as saying, "We'd like to reach 98 percent of the world—that is our goal." Lowe inquired if she was trying to lure her husband to 23andMe. "We're doing our best to poach him," she replied playfully. "We've even offered to double his salary to $2." Ever the supportive spouse, Brin was spotted around Davos proudly sporting a 23andMe "I spat!" badge. For attendees, the opportunity to be a personal-genomics pioneer, one of the first people to glimpse their genetic code and get clues about their genetic predispositions, disease risks, and ancestral origins, proved utterly irresistible. The fact that this opportunity was offered by a trendy firm with personal and professional ties to Google—Brin had personally invested in his wife's venture—just upped the cool quotient. Suddenly, the phrase "Googling your DNA" took on an entirely new meaning.

On November 18, 2007, two months before the sortie to Switzerland, Avey and Wojcicki had introduced the media to the "first personal-genome service." The claim, prominently displayed on the 23andMe Web site, conveniently ignored the fact that the Icelandic firm deCODE Genetics had launched a similar venture just seventy-two hours earlier. Still, Avey and Wojcicki talked about building a new kind of company based upon the achievements in genetics over the previous decade: the declared completion of the Human Genome Project (HGP) in 2003; the ensuing program to catalog millions of DNA variants in the genome; and the explosion of discovery in pinpointing the genes associated with dozens of mankind's most common diseases. 23andMe would bring the fruits of the latest research within reach of the average citizen. The company—displaying its Internet savvy—would also build a dynamic social-networking community that would not only prove fun for their clients, allowing them to share information, track relatives, and so on, but also become an invaluable collective resource for medical research.

Among 23andMe's governing principles were that

- Individuals should have the right to search their own genetic information;
- Individuals should control their own information but can share it with others if they so choose;
- The value of a person's genetic information will increase over time;
- Privacy is paramount.

23andMe clients initially paid $999 for the privilege of spitting a few teaspoons-worth of saliva into a sterile tube. Once the barcoded sample arrived at a CLIA (Clinical Laboratory Improvement Amendments)-certified lab, the DNA was extracted, processed, then passed into a device called a DNA microarray made by Illumina that scanned the identities of about 600,000 specific bases (about 0.02 percent of the 3 billion bases) in that person's genome, revealing which base (A, C, G, or T) resided at each location. A small but rapidly growing subset of those single nucleotide polymorphisms (SNPs) is associated with common diseases. The genotyping results were then sent to 23andMe's headquarters in Mountain View, California, where scientists assessed the impact of the variants on the risk of various diseases. After a few weeks, an e-mail informed the client that his or her results were available via 23andMe's secure Web portal. This was a crucial part of the company's philosophy of empowering individuals, presenting an individual's DNA results in the context of continuously updated and validated information on select genes and traits. The proprietary software used by 23andMe allowed individuals, in Wojcicki's words, "to gain deeper insights into their ancestry, genealogy and inherited traits and, ultimately, the option to work together to advance the overall understanding of the human genome."

The name 23andMe was Wojcicki's idea. My own slightly embarrassing reaction upon hearing it for the first time was to ponder a possible tie-in with LeBron James. "We wanted something that was going to be 'science-y' but not necessarily so nerdy that people weren't going to understand it," Wojcicki told me.[3] "I was at home reading about chromosomes on Wikipedia," which sounds fairly nerdy. "I saw they had a nice picture of the 23 chromosomes, and I was like, 23 . . . and . . . *me*. My husband was in the kitchen and he was like, 'Oh yeah, 23 and me.' " Avey, not, she admits, the most creative type, thought the name was a breath of fresh air. The pair tend to pronounce the name with an emphasis on the "three," as if they're underscoring that their company is about the human genome (which has 23 pairs of chromosomes), as opposed to, say, chimpanzees (24) or rabbits (22).

The launch was beautifully choreographed with judiciously timed stories in the *New York Times* and *Wired* and the rapt attention of a fascinated blogosphere.[4] At the firm's first spit party, Nobelists mixed with movie stars, including Goldie Hawn and Kurt Russell, while 23andMe staffers modeled T-shirts bearing the slogan "Great Expectorations." About the only marketing misstep was that, for $999, 23andMe didn't see fit to include the cost of shipping—FedEx charges were extra.

Not all the attendant coverage of the birth of consumer genomics was favorable. One blogger, listing "The Dumbest Start-Ups of 2007," wrote, "For those who think that spending that type of money while giving up ownership of your saliva is not a winning proposition, [we] can advise you that the results of your 23andMe analysis would not have been promising anyway: you will one day die."[5] And Helen Wallace, director of the pressure group Gene-Watch UK, worried, "The human genome is set to become a massive marketing scam . . . there is a real danger people could be misled about their health."[6]

But this didn't seem to faze 23andMe one bit. Shortly after Davos, 23andMe bagged its first network television profile on ABC-TV's *Nightline*. While host Martin Bashir nervously waited to learn his genetic predisposition to obesity and diabetes, Pixar executive Greg Brandeau (a former colleague of Avey's) said he tried 23andMe because his wife was Chinese-American and he wanted to learn about the genetic makeup of his children.[7] But for Avey, some of the most rewarding moments had nothing to do with the celebrity clientele. She and Wojcicki were taping an interview for CBS-TV in New York when a cameraman approached them. "I'm adopted," he said. "I don't care how much this costs."[8]

The media blitz continued with a live appearance on NBC-TV's *Today* show, with correspondent Peter Alexander serving as the genomic guinea pig,[9] admitting he had been reticent about sharing his DNA on national TV. Wojcicki ticked off the highlights of "Peter's Genes," including DNA variants that determined his blue eye color, ability to taste the bitterness of Brussels sprouts, and wet earwax. When Alexander revealed that he carried seven high-risk markers for type 2 diabetes, Wojcicki had a positive spin: "If you start inputting information about yourself and we start following you, my God, in five, ten years, we could have really powerful genetic information that is going to be able to change the research world." Next came "Peter's Roots," which he deemed the coolest part of the exercise. For this, 23andMe studied variants in Alexander's mitochondrial DNA (passed exclusively through the maternal line) and his Y chromosome (from father to son). While his mother's maternal relatives came from Europe, his father's roots were in Algeria and Saudi Arabia. "They were likely from Bedouin populations, which may indicate why I like to travel around the country doing this for a living."

In the summer of 2008, I visited Avey and Wojcicki at 23andMe's original headquarters (it's moved since then) in Mountain View, California, about forty miles south of San Francisco, in a nondescript two-story office on the fringe of the sprawling Google campus. A bulletin board in the reception area proudly displayed press clippings and magazine spreads. A couple of large glass jars contained 23andMe badges. One depicted two stalks of asparagus and the slogan, "I can smell . . . when I pee!" While I waited, a handful of employees walked upstairs to take a yoga class.

I felt a slight sense of apprehension at meeting the spouse of a billionaire for the first time, but Wojcicki greeted me with a cheerful wave as if we were old friends. (She had met Sergey Brin after the Google founders rented the garage of her sister's house in 1998; the two were married on a sandbar in the Bahamas in 2007.) As nonchalantly as possible, I inquired about her reasons for launching the company. She began investing in health care companies, she explained, after she graduated from Yale in 1996 with a biology degree. "I loved it initially," she said.[10] "You had all these companies that were taking on passionate innovation around genetics. You had 'The Race' for the Human Genome Project." She ticked off biotechs involved in the genomics gold rush, like Incyte, Millennium, and Human Genome Sciences. But once the dot-com bubble burst in 2000, the investment market shifted into "me too" mode. Companies leveraging the genome either expired, were acquired, or tried to reinvent themselves. Wojcicki's enthusiasm for the business waned, not because the investing wasn't fun but because there was no innovation. As she said, "I got into this because I was fascinated by the innovation and the creativity."

In a talk at Berkeley a couple of weeks later, Wojcicki put up a slide that gave the main reason she started 23andMe.[11]

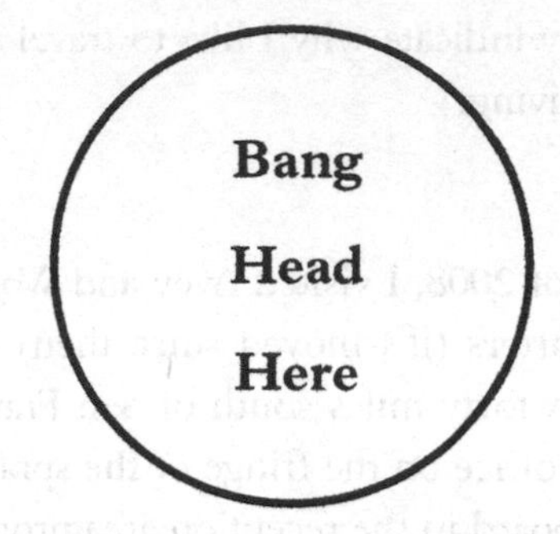

She ticked off the appalling problems bedeviling the pharmaceutical industry, including the abysmal return on $60 billion of annual investment in research and development, the decade or more required to approve a drug, and the industry's stagnant approval rate. But then, she told me a turning point for her came during a biotech conference, when she found herself at dinner sitting next to Markus Stoffel, a Swiss geneticist then at Rockefeller University.[12] Stoffel told her about a project with Jeffrey Friedman—the scientist who famously discovered the "obesity" gene in mice—to study the genetics of a few thousand inhabitants of Kosrae, a Pacific island 2,000 miles northeast of Australia. Due to dramatic changes in diet after World War II, most residents of Kosrae are obese and many develop type 2 diabetes. The island's isolated population was an oasis to tease out the putative genes that predispose to a constellation of symptoms and cardiac risk factors known as syndrome X. Wojcicki thought the project was slightly crazy but remembered Stoffel saying, "I have so much data I can't handle it, but I don't really have enough to make sense of it." That got her thinking about the medical potential of building comprehensive databases of genetic information.

Wojcicki launched into an account of one of her pet peeves, talking faster and even more passionately. "The AIDS community had really successfully mobilized, they're the only community that has genotyping as a routine part of care," she said. "It was unbelievable to me that that wasn't happening in cancer." Pharmaceutical companies such as ImClone and AstraZeneca, among others, could have been genotyping cancer patients to understand why some responded to drugs and others did not. "It's a crime to me that genotyping is not routinely a part of diagnosing cancer," Wojcicki said, barely pausing for breath. Why weren't pharmaceutical companies building databases of information on patients in their drug trials?

Wojcicki credited as her biggest influences 1990s biotechs such as Incyte and Genaissance, which had amassed large databases of gene-based information, and India's Apollo Health. "There are lots of companies that tried it, but nobody ever took on the consumer approach with it," she said. "There's nothing worse than you go to a [cancer] meeting . . . and a patient stands up and asks a question, and it's like, 'You're riff-raff!' Nobody said, 'We're going to have you pay and be a part of this.' I think that was unique. Hopefully we're being innovative."

Wojcicki wanted to empower individuals to gain access to their personal information. Of course, people were not going to be inspired by Big Pharma or the government to pool genetic information, form communities, and revolutionize health care. But in Wojcicki's world of Silicon Valley, Web sites such as YouTube and MySpace were quickly establishing the extraordinary value (and potential profitability) of building online communities from the bottom up. It was only natural that the fiancée of one of the founders of Google should think big. Wojcicki wanted to start a health care revolution—massive online access to genetic information, forming communities that could drive discovery.

That revolution might have started sooner had Wojcicki and Avey not missed several opportunities to meet before their paths eventually crossed. Ironically, Avey met Brin before she met his wife. After working for a string of life science companies including Applied Biosystems and Molecular Dynamics, she was working for a Bay Area genomics company called Perlegen Sciences. After securing a major grant from the Michael J. Fox Foundation for research on Parkinson's disease, Avey tried to arrange a meeting with Google, but the holdup was not so much Brin's schedule as his partner's. "I was like, who is this girlfriend and why does she need to come to the meeting?! I was very confused," said Avey. Brin was advising the foundation on how it was spending its donations. "It was like the stars aligned," Avey recalled. "He came to one of our meetings and immediately went to the heart of what we're doing. What are the algorithms? What statistical rigor do you apply? It was very obvious he was (a) very intelligent, (b) very curious, and (c) I wasn't the right person for him to be talking to!"

Meanwhile, Wojcicki was trying to arrange a meeting between Google and Affymetrix, the leading producer of DNA chips (the popular devices used to study gene structure and activity). But while she got Google excited about a meeting, Affymetrix passed. As fate would have it, Avey later joined Affymetrix, pursuing potential diagnostic applications for the DNA chips. She was jogging with some colleagues when she heard about Google's interest. "Are you guys insane?" she told them. "There's a huge fit. Affy generates a lot of data, Google organizes data, there's got to be something to this." Her colleagues laughed. "Okay, Linda," they said. "Go run with it." With Affymetrix and its chief rival Illumina able to screen hundreds of thousands of SNPs using their DNA chips, Avey thought there could be a consumer focus for the technology, and told Affymetrix that she wanted to bring this genotyping technology to the consumer world. "The prices had dropped to a point where we could bring these large arrays to the consumer market at a price under $1,000," she said.

Linda Avey hardly needed a DNA scan to appreciate the rich values instilled in her family. Her great-grandmother left Germany for the United States at the age of twenty-one, settling in Iowa in 1907. She had ten children and lived to a hundred. "Sometimes when I look at my big German hands . . . I think of the endless chores she performed with hers on their family homestead," Avey wrote on the Huffington Post Web site.[13] Her grandmother, the eldest child, suffered from a cleft palate and bipolar disorder but lived well into her nineties. Avey's mother taught her daughter "an appreciation for blazing sunsets, bright stars on a Midwestern summer night, the song of the meadowlark, the pungent aroma of crushed wild thyme and the precious shape of the Kinikinik flower." Avey confessed that she was not a "typical" mom to her own children, "but maybe it's the mitochondrial DNA we all share that helps us make the best of what we're given."

Avey was aware of at least one company that had tried before to bring genotyping to the general public. In 1998, physician Hugh Rienhoff, Jr. set out to form his own genomics company, DNA Sciences, using a novel DNA-sequencing instrument and aiming to recruit patients to offer their DNA for research. "Academics hoard their patients," he said. "I decided I'd just go direct to the patient. That's when we decided to launch DNA.com."[16] The company will go to the patient's house, collect his or her blood and a medical history, and perform gene association studies for specific diseases. A person might be a control in one study, a test subject in another, and would never be told the results.

The DNA.com Web site launched on August 1, 2000, just five weeks after President Clinton's genome project declaration, promising volunteers that the genetic insights they contributed could change medicine forever. "This is the first opportunity for Six-Pack Joe in the heartland who has diabetes to participate in a study," *The New York Times* quoted Rienhoff in a front-page story.[17] The response was overwhelming, and DNA Sciences struggled to handle the thousands of incoming samples. The more serious the disease, it seemed, the more likely people were to participate. In January 2001, Rienhoff filed papers for an initial public offering, but then the dot-com bubble burst, funding dried up, and in the fall of 2001, Rienhoff resigned. DNA Sciences dissolved 18 months later.

Avey was also encouraged by the growing popularity of the IBM Genographic Project. The National Geographic Society had approached IBM with plans to create a definitive DNA database of global genetic diversity, but one that would be used exclusively for historical purposes, not for medical or pharmaceutical research.[14] The five-year, $40 million "moonshot of anthropology" would map humanity's 100,000-year trek out of Africa and across the planet.

The general public could contribute by buying $99 DNA kits and submitting a cheek swab by mail to gain a glimpse of one branch of their family tree. Kit proceeds supported cultural preservation programs for indigenous populations, with some 250,000 kits distributed since 2005.

Avey had in mind a new personal-genetics company, but who would fund it? And who would provide the computational and Internet savvy to introduce this brave new world to the consumer? The missing piece came in the form of *The Google Story*.[15] The last chapter of the book by David Vise and Mark Malseed recounted a meeting in early 2005 between Brin and Venter at the annual Billionaires' Dinner in Monterey, California. Venter was taking a break from his around-the-world sailing expedition to catalog the oceans' microbial diversity. Having built the world's largest civilian supercomputer at Celera, Venter was acutely aware of the importance of high-performance computing and the desire to "Google the genome." Among the other guests that evening were Ryan Phelan of DNA Direct, and—Avey read with interest—Brin's fiancée, Anne Wojcicki, who entertained her companions with genetics trivia, such as whether they could smell asparagus in their urine after eating the vegetable.

Avey knew of Wojcicki as a successful (and hard to reach) health care analyst, but now the pieces began to fit. In December 2005, they finally met over dinner. Wojcicki called it "a perfect synergy . . . in terms of really wanting to create that tangible Web site that was going to be about your genetic information, as well as combining it with what is now 23andWe," a database to gather and leverage consumer health data. They met again two months later on Valentine's Day, 2006. "Let's just build this company and put 100,000 people in the database—let's just do it!" said Wojcicki. 23andMe was incorporated at the end of April 2006.

From the outset, Avey and Wojcicki saw 23andMe as a largely not-for-profit mission. "We really want to do this because we want to benefit people," recalled Avey at the inaugural Consumer Genetics Show in 2009. "We don't necessarily want to view this as a quick flip of a company to make money—that was not our objective whatsoever."[18] Noble sentiments, certainly, but hardly enough to entice ambitious software engineers away from eBay and PayPal! "You've got to be a for-profit company," Avey said, a little apologetically. "You have to give them something to shoot for, something to get really excited about. We don't see any problem with that necessarily. . . . We think we can do well and good at the same time!" And the company definitely needed to attract and hold onto its top talent. Traditional information-technology and database infrastructures weren't going to work. The 23andMe Web site had to depict a person's entire genome and instantly compare it to many others. Security was another key concern. Even the encryption keys were encrypted. Avey called it "a virtual paranoia" within the company.

At the recommendation of Stanford Nobelist Paul Berg,[19] Wojcicki and Avey recruited as their first two employees a pair of freshly minted Stanford PhDs, Serge Saxonov and Brian Naughton, who bears an uncanny resemblance to Ashton Kutcher. One of their first priorities was to identify about 30,000 medically relevant SNPs to supplement those already included on a commercially available DNA chip sold by Illumina, destined to be the initial 23andMe genotyping platform. The pair combed through back issues of *Nature, Science, Nature Genetics,* and the *New England Journal of Medicine,* among others, cataloging what they considered the most intriguing SNPs linked with disease and drug response. Many of the associations of the SNPs to traits or diseases amounted to little more than hunches, based on studies in only a handful of affected families or individuals, although they might become important as gene associations continued to emerge. To Naughton's immense relief, such associations began to appear in droves in 2007, as if someone had flicked a switch.

To appreciate what happened in 2007, we have to go back a few years. As the HGP went into high gear in the late nineties, an idea circulated. Broad Institute director Eric Lander, who directed the largest American genome center during the HGP and wrote the lion's share of the landmark *Nature* paper in 2001, summed it up thus: "Let's collect all 12 million [SNP] variants, write them down in a very large Excel spreadsheet, all the diseases down the sheet, and associate them with genetic disease." In retrospect, Lander admitted, the spreadsheet idea was "certifiably insane," because only a few thousand SNPs had been documented at the time. But by the year 2000, the number had climbed to 1 million, and the idea of conducting genome-wide association studies (GWAS) for complex diseases like diabetes and heart disease became serious.

Whereas the search for single-gene-caused conditions relied on researching affected families, the hunt for complex disease genes required much larger numbers of patients and measuring the frequency of thousands of markers (SNPs) spread evenly across the genome in large numbers of patients and controls. That prospect became practical after the year 2000 with the commercial production of DNA chips from Affymetrix and later Illumina.

In 2002, the International Haplotype Map Project (HapMap) began, capturing the variation across the genomes of 270 individuals from three ethnic populations: Europeans, Han Chinese, and Nigerians. By 2005, the SNP database from this project was up to 1.5 million SNPs; the number currently exceeds 12 million and could eventually top 20 million. The costs of genotyping also continued to fall. In 2001, a single genotype cost $1, but by 2005, the cost had sunk to one cent. As Lander put it, "Suddenly, a crazy idea in 1996 was quite doable. We could make the spreadsheet."[20]

The first paper reporting a GWAS for a complex disease appeared in 2005.[21] Three American teams found an association between a variant of a gene and age-related macular degeneration (AMD), the leading cause of blindness in the elderly, affecting 10 million Americans.[22] A commentator said the findings finally delivered on "one of the promises of the Human Genome Project [to] provide tools for identifying genetic factors that contribute to common, complex diseases."[23]

In 2007, the floodgates opened. Lander called it an *"annus mirabi-*

lis," one of the most prolific periods of discovery in human genetics. GWAS stories dominated the science headlines, reviving memories of the gene rush of the 1990s, except that now researchers were hauling in genes for truly common diseases—diabetes, heart disease, stroke. The journal *Science* got the ball rolling with more genes associated with type 2 diabetes. *Nature Genetics* followed with a trio of papers on prostate cancer, including studies from the prolific Icelanders at deCODE.[24] The blockbuster study appeared in *Nature* in June 2007, spearheaded by researchers at the Wellcome Trust Sanger Institute, the UK's flagship genome center, and scientists at fifty other institutions.[25] It was essentially seven studies rolled into one. Using Affymetrix chips containing 500,000 SNPs, the consortium pinpointed genetic risk factors in a total of 14,000 British subjects, for seven common diseases, including Crohn's disease, type 2 diabetes, rheumatoid arthritis, bipolar disorder, and heart disease. At 23andMe, Naughton said he was pleasantly surprised how easy it appeared to root out disease genes by studying a bunch of heterogeneous Brits.[26]

GWAS had been transformed from a luxury item to an essential weapon in the geneticists' armory. Almost overnight, geneticists were "drinking from the fire hose" of data. By the end of 2007, the number of risk genes identified had more than doubled, including *FTO*, nicknamed *Fatso*, the first gene implicated in general obesity in humans.[27] The GWAS gold rush continued unabated in 2008, with gene associations found for height and skin pigmentation, LDL cholesterol and waist circumference, autism and bipolar disorder, nicotine addiction and panic disorder, drug and alcohol dependence, kidney stones and narcolepsy, menarche and menopause, and hundreds of others.[28] Researchers studying 8,000 patients hauled in a record thirty-two susceptibility genes for Crohn's disease, like fishermen swapping their rods and reels for large nets. Taken together, the plummeting cost of genotyping and the ensuing GWAS bounty made the personal-genomics era possible.

In March 2007, Google executives visited the lab of Nobel laureate Craig Mello at the University of Massachusetts Medical

Center. Mello told his guests, "Someday, you'll log onto 'Google Genome' and compare your own genome with other people's and maybe get information about what you should be doing and eating."[29] He could almost have been describing 23andMe, which was about to secure its first round of financing two months later. Funding came from two Bay Area institutions, Google and Genentech. Not surprisingly, Wojcicki elected to play down the Google connection, simply noting that her husband's firm was "very good at organizing the world's information. To them, genetics is just another form of information."

Avey and Wojcicki laid low in the run-up to the launch of 23andMe, focusing on making the Web site fully operational and building up the software and scientific teams. A key recruit was Joanna Mountain, a Stanford population geneticist, who had her doubts about 23andMe's business model when she first interviewed in mid-2006. "Clearly these guys had more vision than I did at the time. The science wasn't there yet at all," she said.[30] Mountain signed on in March 2007 as scientific director, leading 23andMe's genealogy work.

The public's fascination with existing genealogy services such as African Ancestry and Oxford Ancestors was hugely encouraging. "If we could get people to pay not just for Y chromosome and mitochondrial DNA but also the rest of the genome, we'd have an incredible resource in terms of understanding human genetic variation. That was really exciting," Mountain said. Her team pushed the envelope in terms of analyzing genome-wide data, helping clients to learn as much as possible about their ancestry, such as where within Asia or sub-Saharan Africa their ancestors came from. Joining the effort was Mike McPherson, another Stanford alumnus, who created 23andMe's groundbreaking ancestry-painting feature. This program parcels an individual's genome into a series of tiny windows and compares the identities of dozens of contiguous SNPs in each window to the millions of cataloged HapMap DNA variants that vary in frequency in Europeans, Asians, and Africans. The 23andMe team then looks at the ratio of the likelihoods that the genotype came from each of the three populations. If the ratio is high, it makes an assignment. If not, then they repeat the analysis, looking at neighboring windows, and fill in the gaps.

As they prepared to launch, Avey and Wojcicki envisioned a service where someone would look at the *New York Times* with their morning coffee and say, "Wow, there's a big story on Alzheimer's disease. How does that affect me?" They log into their 23andMe account, where they could learn about the latest research. The initial service would involve only genotyping, but within a few years, when the time and cost were right, a relatively superficial $1,000 SNP scan could be replaced by a complete $1,000 genome sequence.

In some ways, 23andMe was just another idealistic Silicon Valley software start-up, albeit one that was sitting on top of a trove of genetic data. Aside from allowing people to learn about their ancestry and risk of various diseases and to scrutinize their genetic quirks and variants—what Wojcicki called the "long tail of genetics"[31]—there was another key component. "Our mission is to further research," Avey told me.[32] "The idea is that people would be able to come together, provide a social network, people can tag themselves and say, 'I have autism, Parkinson's and whatever, maybe I'm double-jointed or can roll my tongue.' " 23andMe wanted to build Facebook-style social networks, communities of people with similar traits who could compare genetic information, and, if they desired, share those data with the research community.

While customers could learn a lot about their own genetic susceptibilities to common diseases, 23andMe was treading very carefully. Avey was adamant that 23andMe's service did not constitute a true diagnostic test—even though it was offering consumers the most comprehensive analysis of their human genome so far. The DNA samples were anonymous, bereft of any corresponding or identifying physical data, medical records, or family history, so 23andMe was incapable of producing an individual medical diagnosis even if it wanted to. Writing in 23andMe's blog, *The Spittoon*, Erin Cline put it like this:

> 23andMe is not really a test for anything—not paternity, not ancestry, not disease. Saying our service is a "test" would imply that there is some question you or we are trying to answer. But that's not the point of 23andMe's service. You send your DNA to our contracted lab, they genotype it, and we give you the data—as well as access to what the latest scientific research has to say about it. There is a lot you can do with this data and we're here to help you. But no one should think we're giving them definitive answers to questions like "When will I die?" or "Who's my real dad?"[33]

23andMe chose Illumina to supply the genomics hardware, the platform to scan clients' DNA. Despite Avey's ties with Affymetrix, both founders knew Illumina's CEO, Jay Flatley. Moreover, Illumina's $600 million acquisition of Solexa had provided a bridge to potentially offering whole-genome sequencing of individuals once that was economically and scientifically worthwhile. Three months before 23andMe launched, Flatley gleefully revealed the collaboration during an earnings briefing, suggesting that consumer genotyping would ultimately become Illumina's largest market. Two weeks later, at an investors' conference,[34] Flatley elaborated on the future of consumer genotyping: "We don't think there's going to be people doing this in their garages to genotype their family. We really think this will be a service-oriented business. Samples will come to Illumina or other service providers and the genotyping will be done centrally." Illumina's best-selling array carried 550,000 SNPs and cost less than $450, and a new chip was in the works with double the number of SNPs.

A few weeks later, Flatley divulged that he'd been looking at his own genome highlights, courtesy of 23andMe, making him perhaps the first person to discuss publicly his over-the-counter personal-genome scan. A self-described geek, Flatley reported that

Illumina engineers had developed a genome browser for his new iPhone.

After months of delays, 23andMe finally launched in November 2007. The company had no CEO and only one board member besides the two cofounders: investor and Silicon Valley mainstay Esther Dyson. Wojcicki explained 23andMe's mission in a cover story in *Wired* magazine, playing up the social-networking angle. "It's not that we're against the medical research side of things," she said. "On the contrary, that is where our long-term interests lie. But the genetic profiles are in their preliminary stage and it's too early to be making predispositions and alarming people with the findings. It's the social-networking side of things that will help the product take off."[35] Indeed, 23andMe saw the potential to bring together as many people from as many different populations as possible to forge a new kind of social-networking community. Subscribers could keep their data private but they also had the option of becoming part of a much larger community, sharing their genome data with relatives, friends, and strangers as easily as they might share photos and personal proclivities on Facebook. They could also contribute useful data to the biomedical community. "The power of We," is how Wojcicki hailed the enterprise on the company's blog. "We're at the beginning of a revolution that combines genetics and the Internet. Wikipedia, YouTube, and MySpace have all changed the world by empowering individuals to share information. We believe this same phenomenon can revolutionize healthcare."[36]

In April 2008, I invited Avey to Boston to deliver a keynote talk at an information-technology conference, where she was introduced (at her request) to the beat of "Who Are You?"[37] She reported that customers were already identifying themselves according to their genotype: "When you go to a bar, the pickup line could change from, 'What's your sign?' to 'What's your haplogroup?' " Avey herself had made dozens of online connections with people sharing her mitochondrial haplogroup (H3). 23andMe's ancestry-painting feature allowed customers to get a pictorial representation of their entire genome. Three colors represented chromosomal regions of

European, African, or Asian descent. Avey said her genome was "really boring—all solid blue." By contrast, the mosaic genome of an anonymous African-American male had segments colored navy for European ancestry and green for African. This individual, Avey explained, was two-thirds European and only 33 percent African. "One could maybe argue that this is what Barack Obama looks like," she commented.

A more dramatic example of ancestry painting belongs to Roy King, an African-American psychiatry professor at Stanford and a 23andMe advisor, who wanted to see how his genome compared with his sense of identity.[38] King's maternal relatives hailed from Nashville, Tennessee, and parts of Georgia. Since the end of slavery, many of his relatives had gone to college to become ministers or teachers. In the early 1900s, his great-great-grandfather, a former slave who became a Methodist preacher, wrote about his experiences during the Reconstruction period. King was aware of a Native-American ancestor several generations back; his paternal ancestors were more of a mystery. His father was adopted; his paternal grandmother was African-American and supposedly met King's grandfather, a European-American, and got pregnant, but Virginia law barred miscegenation. There was also evidence of a Jewish bloodline on his father's side.

Growing up, King fully identified himself as African-American, from his appearance to his experiences leading up to the civil rights movement. Family stories of European relatives and admixture did not affect his self-concept, his sense of ethnic identity. But the 23andMe ancestry painting shocked him for two reasons. A slim majority of his genome, 51 percent was blue, indicating European origin. Another 10 percent was of Asian origin, corresponding to the Native-American component of his genome—higher than he expected. "Having been inculcated as an African American, this prompted some reflection. I'm African-American socially, but genetically, am I European-American? How does it affect disease risk for hypertension, prostate cancer, and so on?" Indeed, understanding ancestry is not just for self-identity; it's critical for understanding disease risks. King observed different results for conditions such as prostate cancer depending on whether he selected Africa or Europe for his ethnicity. For now, he must calibrate between the two.

23andMe created an index of personalized genetic information relating to various conditions and disorders (called Health and Traits), which quadrupled in size in the first six months. "People like to paint 23andMe as a fun and frivolous company," Avey said a little defensively, but, she pointed out, her editorial team vetted every relevant research paper to let clients know how much confidence to place in any set of findings. It devised a four-star ratings guide for each condition, like a newspaper movie-rating system, and some papers published in leading journals such as *Science* were assigned just one star because they didn't meet all of 23andMe's validation criteria.[39] For a complex disease such as type 2 diabetes, 23andMe presented data on nine genes that influence risk either up or down, which clients could view to see how each gene contributed to their overall lifetime risk. One caveat, as King discovered, is that the calculations are more accurate for people of European rather than African or Asian descent, reflecting the primary ethnicity of subjects in most research studies. 23andMe hoped to increase the diversity of subjects and help make the calculations applicable to everyone, regardless of race or ethnicity.

Avey cautioned that it was premature for such personal information to become the foundation of too many results that are "actionable." "It's really, really early for us to be making big leaps of faith into what are the clinical end points for this information," she said, until more gene associations are ratified. But she stressed the extraordinary opportunities in social networking in the research space, just as start-ups such as PatientsLikeMe and Sermo were beginning to build online communities for patients and doctors alike. Avey called this budding phenomenon Research 2.0, where cancer survivors and other patients can share experiences and information online. Down the road, some of those data could potentially be sent back to the pharmaceutical industry with a question: "What are you guys going to do about this?"

Avey said 23andMe would preserve the sanctity of the consumer's personal information. "We're here as caretakers of their data. The data belongs to them, it doesn't belong to us." A project called 23 Diseases assigned diseases as priorities for its research, including Parkinson's. "Some people say there's no way you'll ever get self-reported information," said Avey. "We beg to differ." Said Wojcicki, "Linda and I have been criticized for being a little bit too idealistic, but we really think we can change the world. . . . We really think that we can change health care. . . . It has to change, and that's all that we're about."[40]

A few weeks later, Avey and Wojcicki paid a visit to Google to sell some DNA kits.[41] "What we're doing today probably would have cost $500,000 just five or seven years ago. You guys are getting a great deal today at $499!" said Wojcicki. She then had a little interactive fun at the expense of Brin, Page, and Google CEO Eric Schmidt. Based on 23andMe's genomic analyses of Google's Big Three, Wojcicki asked the audience to guess the answers to some Google genome trivia. "Whose paternal ancestor came from Doggerland?" The answer: Page. "Who is 1 percent Asian?" Brin. "Who shares paternal ancestor with Thomas Jefferson?" Brin again. (Wojcicki said the nose was a bit of a clue.) "Who is less sensitive to bitter taste, and thus likes Brussels sprouts?" Schmidt. "Who do you want on your sprinting team?" That was Page, who had a genetic propensity for favoring fast twitch muscles. "Who can't digest dairy?" Brin. "Who can smell asparagus in their pee?" Luckily, they all can.

Wojcicki has divulged some of her own genome highlights. Being of Polish Catholic and Russian Jewish descent, she is especially curious about genes associated with a group of rare disorders that are relatively common in Ashkenazi Jews, including Tay-Sachs disease, Niemann-Pick disease, and Bloom syndrome. After her sister found that she was an asymptomatic carrier for Bloom syndrome,[42] a rare, recessively inherited, cancer-predisposing disease, Wojcicki checked her own results on the 23andMe Genome Explorer. "I typed in 'Bloom's' and saw that I'm identical to my sister on that gene," she said. (Just to be sure, she undertook a separate test using the extended Ashkenazi panel to confirm that she was indeed a carrier.)[43] Her risk of type 2 diabetes was slightly below average. She had a genetic risk for nicotine dependence, which jibed with the fact that she enjoyed smoking when she lived in New York. And she revealed she was a fast caffeine metabolizer. She noted that a small percent of the population abhors cucumbers. "You can look at your shared genomes to make dinner-party guest lists," she joked to her Berkeley audience.[44] She also noted that her mother-in-law and her great-uncle on her Jewish, maternal side hailed from the same area of Eastern Europe and even shared some DNA.

Wojcicki's fascination with her genetic predispositions was not just an academic exercise. At our first meeting in 2008, she revealed that she was pregnant with her first child. "If I did an amnio[centesis], would I actually genotype the child? I would. . . . I think that would be great to have that information."[45] Wojcicki spoke excitedly about checking her baby for penicillin allergy and other drug reactions. "It's crazy to me that you have a new baby, you don't know what it's allergic to! It's trial and error. I was with a group of moms the other day. We all get nervous around snack time, feed our child a peanut and see what happens."

Avey and Wojcicki played up the recreational aspects of 23andMe testing—broccoli taste, fast twitch muscles, and earwax consistency—because they appealed to so many people, but they found the tag irksome. "We really applaud the fact that people are interested in their genomes for reasons other than just their health," Avey said.[46] "The ancestry work we do, I wouldn't call that frivolous at all." Earwax, Wojcicki added, is something that everyone can relate to. "It teaches something about genetics that everyone can understand and is nonthreatening. When you go in the sunlight, do you sneeze? It has spurred dinner conversations all over the world! Genetics has a bad reputation as being associated with death and dying and disease, and it's not—it's about diversity. Diversity's a really spectacular thing that we want to celebrate."

There was nothing recreational, though, about 23andMe's introduction of tests for mutations in the *BRCA1* gene for breast cancer, which sparked a good deal of controversy. Avey said the goal was to empower individuals with pertinent information they could take to their doctor, nothing more. "23andMe never meant to take on that role" of medical testing, she told me. "We were providing information to people that should become part of their medical record. Maybe it was jarring for the medical community to see data points like that." Avey recalled a client who had suffered other types of cancer who learned he too was a *BRCA1* mutation carrier. "How about that?! You can see these lights start turning on in people's heads. Maybe this is correlated. That guy would never have been tested for *BRCA1*. Even though he was Jewish, a doctor would never have thought to offer *BRCA* testing. So we're missing out on lots of data that is useful."

Nonpaternity is another issue that has been sensitive. 23andMe encourages clients to share their genotypes with friends and family and to download their raw SNP data if desired. But with an estimated 10 percent nonpaternity rate in the U.S., 23andMe had to expect cases would crop up. The issue is broached in the consent form, which states delicately: "You may learn things you may not have expected." Avey acknowledged learning about at least one such case, but she insisted that the family in question had no regrets about the test, even leaving 23andMe a voice mail that said, "We're all so glad we did this." Another satisfied customer told Avey that she had sent in her mother's DNA to 23andMe shortly before she died. "Now I have this information on her that I will never be able to get again," she wrote.

There is no doubt that at some point, when the cost is sufficiently low and the information extracted suitably high, 23andMe will shift from offering consumers a genotyping service to whole-genome sequencing. When I mentioned that rival sequencing company Knome had just slashed its whole-genome sequence price by $100,000, Wojcicki immediately wanted to know if it was refunding its charter clients the difference. The issue wasn't so much the pricing as the data quality. "We're not going to sell someone access to their full sequence at $50,000 or $100,000 if we feel like the data aren't high enough quality," Avey said. "If the data quality isn't there yet, it's a shot in the dark."

23andMe rounded off a spectacular 2008 with a string of PR coups. In September, it slashed the price of its DNA kits from $999 to just $399, leaving its competitors looking hopelessly overpriced. It hosted the most exotic spit party of the year during New York's Fashion Week. To the strains of "Whole Lotta Love," equities traders and captains of industry swapped genetic quirks and susceptibilities. Ivanka Trump shared that she had a very low chance of obesity, which she said made her "exceedingly happy," while on the genetics of tongue curling, Wojcicki was quoted as saying that "[CNN's] Anderson Cooper can do a really complicated four-leaf clover."[47] With Wojcicki fully nine months pregnant, the 23andMe cofounders also appeared on *Oprah* via Skype.[48] ("Please have the baby right now," begged Winfrey.) Wojcicki revealed that her baby had a one in two chance of being lactose intolerant and also a one in two chance of inheriting her husband's Parkinson's gene marker. 23andMe also won *Time* magazine's Invention of the Year for 2008, which didn't sit well with deCODE.[49]

Avey surprisingly decided to leave the company in September 2009 to launch a nonprofit foundation called Brainstorm, a charity focused on Alzheimer's disease, motivated by the passing of her father-in-law. Both she and her husband were *APOE4* positive, increasing their risk of developing the disease, but her goal was to trace the other genetic and environmental factors that explain the disease's pathogenesis. In a memo to the staff after Avey's departure, Wojcicki described the goal of 23andMe this way: "We have created a significant and empowering tool, but we must find new and better ways to promote the value of knowing your DNA."[50] It was a tacit acknowledgment that despite gathering more than 30,000 clients in its database in the first two years,[51] and despite the generous backing of Google, 23andMe was struggling with the same harsh realities of defining a sustainable direct-to-consumer business model as its counterparts.

CHAPTER 3

Everybody Wants to Change the World

My first conversation in a couple of years with deCODE Genetics cofounder and executive chairman Kari Stefansson went quite well, all things considered. "Ah, Kevin Davies. . . . A man of miserable repute!" Stefansson enunciated with relish over his cell phone from the capital of what he calls a "tiny rock in the North Atlantic."[1] Apparently he still hadn't entirely forgiven me for a quote of his that I published (accurately, I might add) in an earlier book, in which he called Craig Venter a rude name.[2] (The men later had a good laugh about it.) I'm no expert on Icelandic culture—or humor, for that matter—but it seems that Stefansson gets a kick out of shocking and provoking people just to see how they respond. Nor does he reserve his scorn for hapless journalists. A few days later, during a deCODE analyst briefing in New York, Stefansson (appearing via satellite) publicly rebuked his cofounder Jeff Gulcher for interrupting him, like a father scolding a bothersome child.

From the moment he left his neurology professorship at Harvard Medical School and returned to Iceland to launch deCODE in 1996, Stefansson has been a lightning rod for controversy. Geneticists, bioethicists, anthropologists, and politicians have condemned him and the Icelandic government for allegedly hijacking the genomes of an entire nation. But with the blessing of the Althingi (the Icelandic parliament) and a large majority of the country's citizens, deCODE built a vast genealogical database, taking advantage of Icelanders' proud tradition of recording family histories dating back to the medieval sagas.

Stefansson can trace his own ancestry back a thousand years to the birth of the country and a legendary folk hero named Egil Skalligrimsson. And he is not alone. Iceland was settled more than 1,100 years ago by the Vikings, who brought with them "the prettiest women" from the British Isles as slaves. For centuries, Icelanders have kept meticulous records of their family history. Because of the occasional smallpox outbreak and volcanic eruption, the Icelandic population has remained small, just over 300,000 people, with almost 95 percent being native Icelanders. deCODE's database carries encrypted information on 140,000 citizens, 40,000 of whom have been completely genotyped. Stefansson likes to teasingly say that his countrymen are willing laboratory animals that might even provide helpful information for treating American diseases.

deCODE's mission was to use Iceland's genetics resources to discover disease genes, develop diagnostic tests, and design new drugs. Buoyed by a $170 million initial public offering (IPO) of shares and a partnership with Hoffman–La Roche worth up to $200 million, deCODE mined the Icelandic gene pool for disease genes and fast-tracked several drugs into the clinic. But drug development is a frighteningly expensive and risky undertaking,[3] one that left deCODE drowning in debt and struggling to survive. "You can argue that our company was founded about six years too early, because the technology to systematically isolate disease genes wasn't available really until a long time after we founded the company," Stefansson admitted in 2008. "We've also had a lot of people in the academic community trying to undermine our credibility. People—I'm not going to name them—insisting that our diagnostic tests are not measuring any real risk, etcetera. There are sour grapes in the academic community. This has been rather costly for us." To add insult to injury, many of the DNA variants discovered by deCODE are also featured in the services of its leading consumer-genomics competitors, 23andMe and Navigenics.

Stefansson admitted that deCODE hadn't communicated its story effectively—which amounts to self-incrimination, as he has been almost exclusively the public face of deCODE. Its premature start meant that deCODE collected debt of more than $225 million. Further damage was done by Lehman Brothers, which held much of deCODE's financial assets when it collapsed in 2008.[4]

It is easy to question deCODE's business strategy but impossible to argue with its success in mining disease genes. Since 2001, deCODE has identified genes that predispose to dozens of complex diseases, including cancer, heart disease, and schizophrenia. In the process, it has shed much light on European ancestry, genome cartography, and the genetics of pigmentation, longevity, and many other traits. "We've made a lot of stellar discoveries," said Stefansson. "We're basically leading the field of human genetics." The widely acclaimed Wellcome Trust complex disease paper in 2007 was to Stefansson "an utter failure," because of earlier deCODE results.

Stefansson claimed that more than 60 percent of the "tidal wave of discoveries" in complex disease genes since 2006 belonged to deCODE, leaving other groups trailing in its wake. "We isolated the first [gene variant] in type 2 diabetes with the biggest effect, the first one in heart attack, the first in prostate cancer. The list goes on." Indeed it does: breast cancer, glaucoma, lung cancer, basal cell carcinoma, osteoporosis, and more. Stefansson said, "We are ferociously working on the genetics of complex traits and turning the discoveries into intellectual property [and] diagnostic tests." The first of those tests, deCODE T2 for type 2 diabetes, launched in April 2007, just twelve months after deCODE first reported the association.

With deCODE developing more diagnostic tests, it was a logical step to leverage its gene data to launch what was technically the first personal-genomics service, deCODEme. "The basic goal, if you look at the original business proposal, was always to use genetics for preventative health care," said Stefansson. "The idea of marketing directly to consumers is something we've been toying with for several years. We did not launch deCODEme in response to anything except in an attempt to flesh out our business model."

deCODEme debuted a few days ahead of its more celebrated California rival. "23andMe followed in our footsteps," said Stefansson. "The only advantage that they have over us is they have a lot of money." Since hosting celebrity spit parties on Iceland was not a terribly practical option, Stefansson starred in a series of promotional videos on YouTube[5] to introduce deCODEme, marking "the first time that the lay public is getting real access to the new human genetics." Wearing his trademark all-black attire, as if to accentuate his shock of white hair, Stefansson neatly summarized the rationale for deCODEme and its direct-to-consumer approach:

> We are always receiving negative information about ourselves throughout our life. For example, I have to look at this ugly face in the mirror every morning when I wake up and brush my teeth. . . . Of course, we're concerned about the way in which people will react to learning that they may have a risk of a disease that they didn't know about before. I would, however, be more concerned about a society that would not allow us to explore ourselves, to study ourselves, to learn about ourselves in this manner.

Stefansson stressed that deCODEme was merely an educational service, affording customers an opportunity to deal with any negative information they might receive:

> We are not providing people with a genetic test. We are only allowing them to compare their genomes to the genomes of those who in the literature have been described as having a risk of a disease. We encourage people not to make medical decisions on the basis of results of this, but we point people to the possibility

> of taking results of this to their doctors, who can then prescribe a validated genetic test to make sure that what they are looking at is a true genetic risk factor.

And he signed off:

> [deCODEme] is a service where people, on their own free will—they even have to pass a rather high economic threshold—are coming to us to learn more about themselves. I think that is laudable and I'm convinced that this is going to serve our society well.

Two months after launch, deCODE ran ads in the online version of the *Wall Street Journal:*

> Know your CODE: Learn more about your ancestry, traits, and health risks with the most comprehensive *Genetic Scan* available. From deCODE Genetics, the scientists who discovered the genes.

Stefansson insisted to me that his direct-to-consumer rivals were more like marketing or dot-com entities than genomics companies. "We are the only ones that do the genotyping ourselves. We handle the quality control ourselves. Navigenics and 23andMe are not marketing their own discoveries or progress. . . . They're mostly marketing discoveries that we have made!" Stefansson said he would not want to be in the position of not being in control of the quality of the genetic data, but it was nothing personal. "I like both Anne [Wojcicki] and Linda [Avey]. They are good people," he said. But while his competition was able to include diagnostic SNPs discovered by deCODE "with impunity," Stefansson issued what sounded like a threat. "That is not going to last long. Pretty soon these patents are going to issue."

I asked Stefansson if the fact that 23andMe included some 30,000 additional custom SNPs on its DNA arrays provided an advantage as more complex disease genes came to light. Stefansson didn't appreciate the inference and erupted: "If you ever again in this conversation dare to propose that 23andMe has the know-how to pick up risk more easily than me, I will hang up, all right? How dare you say that? They are just buying some technology off a conveyor belt. We discovered these genes!" As abruptly as his temper had flared up, it blew over. After a brief pause, he said slyly: "How was this for indignation?"

In some ways, deCODEme fell between the offerings of 23andMe and Navigenics. It offered information on fewer complex conditions than 23andMe but more than Navigenics. deCODEme presented information about ancestry, allowing clients to compare their own genome with representatives from different populations—from an Mbuti pygmy to Stefansson himself—but didn't push the social-networking angle as far as 23andMe. On the other hand, deCODE did offer genetic counseling to customers on demand.

Stefansson bristles when he hears criticism about the supposedly trivial nature of the risk information that deCODEme offers. "Let me now go into how trivial they are," said Stefansson when I dared raise the subject. deCODE's prostate-cancer markers, for example, detect the 1 percent of European men who have an absolute lifetime risk of the disease above 50 percent and a relative risk of more than 3.0. (*Absolute risk* is the actual risk of developing a disease over a certain period, typically a lifetime. An absolute risk of 0.5 means that person has a 50 percent chance of developing that disease. *Relative risk* is the ratio of a disease occurring in one group—say, smokers or a group carrying a particular gene variant—compared with a control group. For example, if someone has a relative risk of 1.5, they have a 50 percent greater chance than the matched control of developing a disease.) By comparison, he said, "If you take the risk factors that our society has been using, for example elevation of cholesterol, those in the top quintile of the cholesterol range have a relative risk for heart attack of 1.6. . . . The risk we are picking up with our prostate-cancer test is measuring many times greater risk than cholesterol." (In fact, the gene test can identify the 10 percent of patients with the same increased risk as those with elevated cholesterol.) This information isn't just important in detecting disease risk—as Stefansson's friend and colleague Jeff Gulcher can attest (see the Introduction)—but could impact which drugs patients should take. Prostate-cancer drugs carry unwanted side effects, making them better suited for men with particularly high risk.[6]

"We aren't marketing tests unless they have greater effect on risk of disease than elevated cholesterol," Stefansson reiterated. "It is the single test that has had the biggest impact on the development of internal medicine over the past twenty-five years. It's the kind of risk that society knows how to use." Stefansson ticked off other examples. deCODEme picks up the 2 percent of women who have a greater-than-50-percent risk of breast cancer, primarily [HER2+] estrogen-receptor-positive tumors that can be prevented with tamoxifen. The heart-attack test identifies individuals who have a relative risk of 1.7 of developing an early heart attack, helping middle-aged men and women seek intervention. And in type 2 diabetes, the test identifies patients who are less responsive to sulfonylurea drugs, as well as prediabetics most likely to convert to full-blown diabetes and who are good candidates for the drug metformin.

Although Stefansson is an MD, he has a very high threshold for seeing his physician. He called it a "defensive posture" because "they invade my privacy in some way." Thus it was important for him to be able to determine his personal risk for diseases independently. He said, "There are some people who buy deCODEme who want to look at the data with a genetic counselor with them. Others want to look at it with a physician, some just with their spouse sitting next to them. And then some strange people like me, who close the door before they log on to the Web site because they want this to be a private thing. I don't think anyone should be forced to have a person with bad breath sitting next to them when they're looking at private information."

I asked Stefansson if he thought that full-blown genome sequencing would eventually replace genotyping. "I think it is very likely that it will," he said, probably between 2010 and 2013. Sequencing would ferret out the rare variants that Stefansson said were particularly important in the genesis of diseases of the brain. He explained in his inimitable manner, "This instrument, this organ that compels our statisticians to play poker every other Thursday night and looks like the ultimate luxury organ, appears to be more essential to reproductive success than any other organ in the body. If you have a subtle variation in the function of the brain, you have fewer children. . . . So there is a negative selection of the variants." For good measure, Stefansson had also found a gene association with a love of crossword puzzles.

Appearing via satellite at deCODE's annual research-and-development day in New York in June 2008 (he was unable to fly because his passport had been stolen at his health club), Stefansson sounded his familiar themes. "In five to ten years, pretty much every college-educated person in America will have tests such as deCODEme," he predicted. Jeff Gulcher, deCODE's other co-founder, said deCODE's intent is to convince physicians to change their management practices and convince insurers that it's cost-effective to screen patients. This emphasis on partnerships with physicians makes deCODE less dependent on direct-to-consumer strategies than its competitors.

deCODE doesn't have the promotional pizzazz of 23andMe, but in February 2009, Stefansson was a game, if rather incongruous, guest on the *Martha Stewart Show*. Stewart had visited Iceland the previous summer and said she was excited to help open a door onto the future of medicine.[7] "Almost every child born into this world is going to have a genetic profile done just to make sure that the individual knows what risk they have, so they can do something about it. For some individuals who have not been careful enough in the selection of their parents, it can be important," he winked. Rather than reveal her own results, Stewart volunteered three of her staffers to discuss their deCODEme scans. One producer had a 33 percent risk of age-related macular degeneration. Stefansson advised him to visit an ophthalmologist and lose some weight because he was "too plump." Stewart's assistant, Liesl, learned she had a one-in-five lifetime risk of breast cancer—60 percent higher than average. Stefansson reassured her that she was still "a beautiful woman with a beautiful genome."

Despite the orgy of successful genome-wide association studies (GWAS) published by deCODE and other groups, the development of personal genomics was being slowed by the fact that geneticists could identify only a small fraction of the total genetic variation associated with any given trait. Perhaps the missing "dark matter" could be explained by very rare variants—less common than a typical SNP—that had relatively large effects. If so, those would be found only by whole-genome sequencing.

The most outspoken critic of what he saw as GWAS hype was Duke University's David Goldstein, whose studies on the Y chromosome and the Lost Tribes of Israel[8] explained in part why the *New York Times* saw fit to cover his anti-GWAS rant.[9] "Most reported associations reflect real biologic causation," said Goldstein. "But do they matter?"[10] Goldstein wasn't criticizing the quality of published GWAS but the value of continuing to run GWAS in larger and larger groups of patients after teams had already identified the supposedly strongest variants. More and more studies were implicating more and more genes with minimal effects, and Goldstein argued that "in pointing at everything, genetics would point at nothing . . . There are probably either no more common variants to discover or no more that are worth discovering."

Dennis Drayna, a geneticist at the National Institutes of Health, sympathized with Goldstein, even though he was waist-deep in GWAS himself. The push to find variants with smaller and smaller effects reminded him of the old joke about the difference between Harvard and the Massachusetts Institute of Technology (MIT). "At Harvard, you learn less and less about more and more until you know absolutely nothing about everything. At MIT, you learn more and more about less and less until you know everything about nothing. We're going to learn more and more about less and less."[11]

Take height, for example. Studies identified at least twenty SNPs associated with human height, but in sum they accounted for only 2 to 3 percent of height variation in the normal population. If common variants were responsible for most of the hereditary component of height or diabetes or some behaviors, then, Goldstein argued, genetics wouldn't be much help in elucidating the biology of such conditions, because there would be a ludicrous number of "height genes" or "diabetes genes." Goldstein believed that the search for the missing genetic components should focus on rarer DNA variants with larger effect. Studies by Helen Hobbs at the University of Texas Southwestern Medical Center showed that targeted sequencing could identify such "Goldilocks" variants.[12] But some scientists thought Goldstein went too far, arguing that the value of GWAS is not so much in predicting personal disease susceptibility as in identifying likely disease pathways. Harvard's Joel Hirschorn pointed out that the path from elucidating the structure of cholesterol to the development of the class of statin drugs took a century and the awarding of three Nobel Prizes. GWAS could facilitate much quicker paths from gene to drug, as deCODE once demonstrated in moving from heart-disease genes spotted by GWAS to a drug in clinical trials in just two years.

At the end of 2009, deCODE finally succumbed to the inevitable and declared bankruptcy. After restructuring, it emerged in early 2010 with an American CEO (Stefansson was now president of research) and a new, leaner mission focused on gene discovery and diagnostics. The firm quietly doubled the price of its deCODEme complete scan to an uncompetitive $2,000, steering physicians toward its more profitable disease-specific diagnostics kits and essentially bowing out of the direct-to-consumer arena. When I next saw Stefansson, he was visiting Boston, ironically one of the few Europeans to make it across the Atlantic during the dramatic eruption of the Eyjafjallajökull volcano. But he was eager to return home, for with a change in wind direction, ash was threatening to fall on his horse farm just outside Reykjavik.

In June 2005, I hosted a modest biotechnology awards dinner at the National Press Club in Washington, D.C., for about 150 guests. One of the winners that night was Dietrich Stephan, a neuroscientist based in Phoenix who had pinpointed a gene underlying a rare inherited form of sudden infant death syndrome (SIDS).[13] Stephan is a charming man who seemed genuinely thrilled to receive a $65 mail-order trophy and to have his research recognized by his peers.[14] I just hoped for his sake that this wouldn't mark the pinnacle of his scientific career. I lost track of him for couple of years, until a press release floated into my in-box announcing the birth of a new personal-genomics company called Navigenics. One of the cofounders was David Agus, a celebrated Los Angeles oncologist. To my pleasant surprise, Stephan was the other.

Stephan's work on SIDS was just a sample of his impressive genetic detective work conducted at the Translational Genomics Research Institute (TGen) in Phoenix. After the completion of the Human Genome Project, Stephan's interest began to shift to more complex diseases, but the flood of GWAS and accompanying media coverage had a downside: "We're having these huge lines of patients walking into physicians' offices saying, 'Well, I just read this chromosome 9p association in the *New England Journal*—give me the test. Tell me what this means about my risk.' "

"Huge lines" was a bit of an exaggeration, but what could physicians do when confronted by an eager, informed patient who'd read GWAS stories in the popular press, besides offer the tired medical platitudes: don't drink, don't smoke, take a baby aspirin, and watch your weight. Stephan's deep frustration over the total absence of diagnostic implementation of genetic risk factors for heart disease, cancer, and Alzheimer's drove him to launch Navigenics. When I called him after receiving the press release to congratulate him, he told me: "Someone has to draw some lines in the sand, start framing up the infrastructure for really applying common variant risk assessment for common human diseases, the ones we're all going to get in our lifetimes."[15] Although he doesn't dwell on it, there was personal motivation as well. His mother died from breast cancer when he was seven years old. "The doctor said, 'You have a lump, come back next week.' By then she was full of metastatic disease." It's that archaic, reactive brand of medicine that Stephan was determined to break.

The aims of Navigenics were written up in a November 2007 story in the *Wall Street Journal*[16] just before the launch of 23andMe and deCODEme. However, the firm's Health Compass service did not get off the ground until spring 2008. After watching 23andMe monopolize the media for six months, the Navigenics brain trust was obligated to make a splash of its own. The company leased a storefront on a cobblestone street in the trendy SoHo district of Lower Manhattan, an area better known for designer boutiques such as Anna Sui and Helmut Lang. Amy DuRoss, Navigenics vice president of policy and business affairs, liked the notion that people were drawn to SoHo by new and exciting trends, even if it meant partying opposite a boutique displaying mannequins in hot-pink lingerie. For DuRoss, planning a fortnight's series of receptions and seminars was trivial compared to her previous gig—coauthoring proposition 71, the $3 billion California stem cell bill that passed in 2004.

On April 8, 2008, I joined some two hundred guests at the launch party, the networking facilitated by lashings of Navitini cocktails.[17] Guest of honor was former Vice President Al Gore, who had just flown in from Iceland and another *Inconvenient Truth* lecture. In a short opening speech, Gore, a new partner with Navigenics's lead investor, Kleiner Perkins, hailed his friend Agus as "a genius in oncology" and "a miracle worker." "This is a great firm," he said. "They've got the ethics and the culture and the values right. . . . On all these new genetic breakthroughs, there is always some resistance culturally, and then, when there's an evaluation of the inherent value, if the ethics are right, if the surrounding culture is right, then it just breaks through. . . . I think it's going to be a fantastic success."[18]

Interestingly, Gore's appearance fulfilled a prophecy he had made a decade earlier, when he was vice president of the United States. Delivering the James D. Watson lecture in 1998, Gore predicted: "Within a decade, it will be possible for our doctor to take a cheek swab, place a few of our cells on a gene chip scanner, and quickly analyze our genetic predisposition to scores of diseases."[19]

Moderating duties at the event were handled by Greg Simon, a former Gore advisor who chairs the Navigenics task force on policy and ethics. In the same vein as the Hubble telescope, Simon said, Navigenics had developed the GeneScope, which would deliver the world of genetics to the public in practical ways to improve their health. "All health is personal," Simon said. "It's only the doctors and scientists who've had to deal with health in an impersonal way, because they didn't have the tools to get personal about the drug in front of them."

Just before he was introduced, Stephan told me that after eighteen months of tireless work preparing to launch Navigenics, all he really wanted to do was go to sleep. But he got his second wind. "Let me get this straight," said Simon, doing his best Stephen Colbert imitation. "I spit, you SNP, you call, I pay. Is that basically what happens?"

"You pay first," said Stephan with impeccable comic timing.

Simon next introduced Agus as the "doctor to the world," which sounded better than proctologist to the Hollywood elite. Agus was just forty-two with cholesterol safely in the mid-190 range and no family history of heart disease. But his Navigenics genome scan revealed an 82 percent lifetime risk. "This was staring me straight in the face," he said. "That 82 percent hit home." Agus went on the statin drug Lipitor and started exercising and watching his diet. "The way we're going to make progress is if people go to their doctor, and say, 'I've had this test done, this is what it says. Let's do something about it,' " said Agus. "We spend 60 percent of our health care dollars in the last two years of life—that sucks!"

As the guests filed out of the event, Navigenics staffers handed out custom T-shirts festooned with genetic slogans and $500 discount coupons for the Health Compass. Everyone grabbed a shirt, but I didn't see many people placing orders for their genomes. Doubtless the $2,000 price tag (including discount) was a deterrent, but there was another problem. Beside the stacked boxes of saliva-collection kits there was a sign informing residents of New York State that they couldn't receive results until the state's Department of Health officially approved the test. Navigenics assured potential buyers that they could still order the tests without their credit card being charged until such approval was forthcoming.

Walking around SoHo with a pleasant Navitini buzz, I found a cozy Italian restaurant and ordered a late supper. A friendly South African waitress named Suzaan spotted my new Navigenics tote bag, and we started chatting. Suzaan's extended family was riddled with ovarian, breast, and other cancers. She'd moved to New York from Cape Town nine years earlier at the age of twenty-one, having taken the $3,500 BRACAnalysis test from Myriad Genetics. Her results came back positive—she was a carrier of the *BRCA1* mutation, increasing her chances of developing breast or ovarian cancer. Ever since, she told me, she's religiously undergone a mammogram every six months, watchfully waiting for the seemingly inevitable moment when, in the words of the late British journalist John Diamond, "a few of these cells would band together in that state of cytological anarchy which leads to cancer and death."[20]

"I'm the luckiest person in the world," David Agus, a.k.a. Doc Hollywood, tells me as I am ushered into his cozy Beverly Hills office by his charming receptionist, Autumn, not surprisingly a theater major. There were no photographs that I could see of his famous Hollywood clientele, and as he sat in front of his three flat-panel monitors, he looked more like a day trader than a world-renowned oncologist. He was of small build with curly hair, but spoke with a quiet intensity that comes from trying to save lives every day—not always successfully. "I get to help people, and at the same time I get to do research. Half of my job is bringing hope," he said in explaining why he launched Navigenics:

> A clinical trial or a new drug brings hope to an individual. The reality is, it's only going to work in 10 to 20 percent of people. But that hope is what keeps people going. I did Navigenics mainly because I'm sitting there seeing people dying of cancer. The notion of curing cancer is, to me, somewhat way out there. So I want to prevent it. I can get a lot more mileage by pushing that prevention side.
>
> If you come with me to see a patient here, I'll look and hear all the data and I'll talk to them, and together we'll make a decision. There's no right decision. It's not a binary yes/no. It's the patient's value system. It's the data. It's my heart, having seen 1,000 cases.[21]

One day, however, Agus believes it will come down to this: gene X is up, gene Y is down, here's the pathway that connects them: "The notion of calling diseases by body part is a remnant from the mid-1800s in France. . . . It ain't gonna be that way in the future. It's going to be characterized by pathways and signals."

Agus trained at Johns Hopkins and Memorial Sloan Kettering before meeting junk bond king Michael Milken, who provided "a few resources" to persuade Agus to study prostate cancer. The first week after moving to Los Angeles in 2000, Agus, a self-described "Brooklyn guy," received a phone call from a Hollywood producer. "I heard there's a new test for prostate cancer. Do you have the new digital equipment?" Agus held up his middle finger. "Sure—this is the digit," he laughed. He's gotten to know many more Hollywood power brokers since then, none more famous than Steven Spielberg. "Steven's been great," is about all Agus wants to say, adding that he never asked the director for research funding. Agus was preparing to move his team to run a new center for molecular medicine at the University of Southern California. "We're very lucky.

Larry Ellison and Stephen Spielberg and Sumner Redstone and Eli Broad have set up this new institute for us."

The seeds of Navigenics were planted in 2004, when Agus happened to read an article in *Fortune* magazine. "I still remember the day walking from the hospital to the clinic, there's a gift shop there, and in the window was *Fortune* and the headline 'Why We're Losing the War on Cancer.' "[22] The story by cancer survivor Clifton Leaf left a deep impression on Agus. Leaf pointed out that the cancer death rate hadn't changed in fifty years. "We've had this very reductionist view of all biology," says Agus. "We go down to the cell, the pathway, and we start to forget the big picture, why we're doing it." Leaf's story left Agus both despondent and hopeful.

A little later, attending a National Cancer Institute retreat in the Arizona desert, Agus met his friend TGen director Jeff Trent for dinner at a Mexican restaurant. Agus told Trent, "We need to apply this stuff to patients. I know it's early, but we've got to show this stuff works or else it's not going to take off." Trent said one of his colleagues was thinking along the same lines—wanting to apply genetics to individuals. "We should get the two of you together." It took just five minutes on the phone for Stephan and Agus to realize they shared a vision. Backing came immediately from noted venture capitalist John Doerr, who was on the board of ABC^2 (Accelerate Brain Cancer Cure), a foundation that Agus helped launch following the diagnosis of the late Dan Case, brother of AOL founder Steve Case and chairman of Hamblett & Quest, with advanced glioblastoma.

In contrast to Agus's exuberance, Stephan's calm demeanor belies a fierce determination to improve health care. For fifteen years, his research was predicated on the belief that all human disease has a genetic component. The earlier the intervention, the better the outcome for the patient and the greater the likelihood of developing therapeutics, and Stephan wasn't willing to wait another twenty years until all the relevant gene-gene and gene-environment interactions were teased out before offering some clinical implementation. "People thought it was really healthy to smoke. So why did it take us five decades to put the stamp on the cigarette package that smoking is dangerous to your health? Did we really need 50 years of epidemiology to put the sticker on? I would argue no." The medical message of the twentieth century, according to Stephan, can be boiled down to: "Don't smoke, don't drink, and don't be fat. And don't let your kids eat paint chips. That's 100 years of epidemiology. It begs the question: Are we over thinking this stuff—imposing rigor on the system that isn't there?"[23]

Everyone has heard of people who smoke two packs of cigarettes a day and live to a hundred. On the other hand, about 10 percent of lung cancer patients have never smoked in their lives. Rather than rely on population data, Stephen argued, it was time to provide individuals with the tools to learn whether their genetic risk factors are increased or decreased relative to the general population, how serious those risks are, and how they interact with the environment. Stephan wanted to put another tool in the physician's toolbox "that's going to go into medicine over the next decade."

Stephan and Agus assembled experts in ethics, legal affairs, regulatory policy, genetic counseling, science, epidemiology, risk communication, and publishing. Their task would be to translate and communicate sophisticated genetic information to a lay audience, including physicians. The highest-quality standards were paramount, so as not to destroy consumer confidence and damage the field. "From day one, we're making sure that every nuance of this operation is perfect," he said. A critical decision was to offer genetic information direct to the consumer as a way to keep the information private and confidential.[24] "Given that [a] person owns their genome and no one has a right to it, in my opinion, that's how you have to start—by giving it to them and them alone," said Stephan. "Your genome belongs to you and you alone." But Navigenics also insisted that its service did not constitute a diagnostic test or medical advice, and disclaimers warning against the use of this information without appropriate medical consultation were prevalent throughout the company's Web site and materials.

During a visit to Navigenics headquarters shortly before its launch, I sat in on the weekly meeting of the company's editorial team, a group of young geneticists, epidemiologists, statisticians, computational biologists, and bioinformaticians tasked with vetting every major published genome-association study. The group was chaired by Sean George, but the scientific leader was Michelle Cargill—a geneticist with the rare distinction of stints with both Eric Lander and Craig Venter.

Stephan, who suspected that many of the reported GWAS results were tantamount to garbage, attached great importance to the triage process. According to this, regardless of the prestige of the journal or reputation of the authors, at least two PhD team members delved into the results to judge whether the purported gene association met three major internal criteria. First, was there sufficiently dense coverage of SNPs across the genome to detect huge effects? Second, was the association replicated in at least two independent populations?[25] If a reported association in a French population isn't replicated in American subjects, it is likely a false positive. Third, was the study controlled for ethnicity, gender, age, and other risk factors?[26] A good example of this "decision heuristic" followed the Wellcome Trust's blockbuster GWAS paper of 2007.[27] "No disrespect to the Wellcome Trust folks, but their strongest association was in bipolar disease," said Stephan. When the Navigenics team examined the SNP distribution, "we [concluded] the association may have been a false positive. Because of that possibility, it won't be on the Navigenics menu."[28]

Stephan was under no illusions about the complexity of delivering clear, reliable, actionable information to the general public. The science journals, he said, are clogged with false-positive reports—hints of disease linkages and associations that haven't been replicated. What happens when consumers embrace the $1,000 genome, demanding information on long lists of incriminating mutations that have not been verified? "Then you're misinforming people and potentially doing them damage," said Stephan. "If you link to a risk variant associated with ALS [amyotrophic lateral sclerosis, or Lou Gehrig's disease] that has never been replicated, and you don't communicate it well, that person could walk away with the thought, 'Oh my God, I'm going to get ALS when I'm fifty'—and go jump off a building. That's a real risk with this type of information." But that person may simply be a carrier of a variant, like most of the population, with fairly average risk. Communicating the information around both common and increasingly rare DNA variants is the key.

Like its competitors, Navigenics also makes a good effort to communicate the dual role of genetics and environment (including ethnic background) in shaping individual risk of complex diseases. Eventually, Stephan wants to include an "environment calculator" to complement the genetic data, so clients can select and deselect relevant environmental risks, combining all known genetic and environmental risks into one amalgamated risk score: "If you have risk factors for prostate cancer and you eat fatty foods, you'll have increased risk. But we don't yet understand the gene-environment interactions." Stephan offered autism as an example. There might be ten different subtypes of the condition that are susceptible to different environmental risk factors, he said. The ultimate prediction won't happen until researchers carry out an enormous trial of people from different ethnic backgrounds, exposed to various risk factors, drop them into a single model, and, as Stephan is fond of saying, "push a button so you can see every combination of risk factors and how they'd fare if they had that combination of risk factors."

Unlike 23andMe, Navigenics focuses squarely on risk assessment for actionable, common medical conditions such as heart disease, cancer, and diabetes. Navigenics does calculate a customer's ancestry, but that information remains in the background to help refine ancestry-specific risk assessments. Stephan and Agus aren't interested in providing information on fast twitch muscles, asparagus smelling, and other quantitative traits. "We don't want to dilute what we do best by doing these offshoots," said Stephan, who also worries about potential ethical problems, from nonpaternity to informed consent. "Ten percent of people don't have the father they think they have, so suddenly, you're going to have family units exploding because they think it's going to be fun and interesting. Children can't give an informed consent, so some may be genotyped at twelve. Then when they're thirty, they're distressed because they were forced to be genotyped by their parents."

Nor will Navigenics include nonactionable conditions—diseases without a beneficial medical treatment or lifestyle change—because that can be devastating. "Deafness or Down's syndrome—we're not touching them," said Stephan. For example, even if Navigenics validated a SNP for ALS, that information would be included in the Health Compass only if there was a legitimate therapy or lifestyle modification on the market.

In keeping with its emphasis, Navigenics encourages customers to print out their results summary to take to their physician.[29] Customers can also obtain access to their raw SNP data, but before they claim their CD of the data they must first wade through a sobering four-page consent and release from warning that the data might not be interpretable, complete, or accurate; that the data's clinical significance could be misunderstood, leading to unnecessary or delayed treatment or discrimination; that the data could be associated with untreatable diseases; and that "you may learn information from your genome-wide scan data file that could potentially cause you psychological or emotional distress." If you get that far, the result is a rather large genotype file—some 900,000 rows of data. Stephan tried once to print out a file using 8-point font and got to two thousand pages before the program crashed.

Stephan was among the first volunteers to take the Health Compass test.[30] "No one who's taken it is frightened or scared when they get their results back," he told me. People had an intuitive understanding of this kind of testing. As for his own results, Stephan said: "Great news—I don't carry an *APOE4* allele. That has major implications on my stress levels and how I live my life." But he also learned he has an increased risk of breast cancer, which, given his family history, might prompt him to get a checkup (although breast cancer is extremely rare in males). In contrast with Stephan, Agus was unusually reticent. "I'm like every Jewish guy, neurotic about health," he told me. "I don't know how they got the first group of people to do it!" Some years ago, he recalled, Bobby Kotick, the smart, aggressive CEO of Activision, the video game company, held a fund-raiser for Agus called Future Health Day. Kotick had just turned forty, so he gathered nine friends and held a medical testing bonanza—blood panels, heart scans, colonoscopies, the works. "We picked up three cancers," said Agus. "At the same time, we empowered them on their health care. That's what's exciting."

What finally inspired Agus to take the test himself—and learn about his double lifetime risk of heart disease—was a call from a forty-one-year-old woman. Her DNA results indicated an elevated risk for colon cancer, so she had a colonoscopy that revealed a 3-centimeter polyp. Her father had died of cancer, although her mother didn't know what kind. Her story convinced Agus that the testing could have a beneficial effect. While there's no certainty that the polyp would have become cancerous, he was convinced that her risk of cancer was lessened as a result of surgery. In fact, of the hundreds of friends, family, and other volunteers who have tried the Navigenics test, almost half took some positive action after learning their results—joining a gym or visiting their doctor.

Agus shared another story about a woman in her forties who learned she had a more-than-double lifetime risk for Alzheimer's disease. "It was a wake-up call for her—she's not going on a statin, but she's leading her local city Alzheimer's Association. She's out there now, working for a cure. She's saying, 'I feel empowered, I'm doing something to help in the future. It's fantastic.' "

What is the benefit of detecting Alzheimer's early on? Stephan recited the alarming numbers: the prevalence of Alzheimer's is expected to quadruple by 2015 to between 16 million and 20 million patients, costing the U.S. health care system up to $2 trillion annually. "It's simply unsustainable from a health-economics point of view," he said. Detecting Alzheimer's at the age of twenty-one opens up what Stephan calls "the therapeutic window" of maybe fifty years: "Rather than presenting when half your brain is being melted by disease, you present when you're totally cognitively normal."[31] With emerging therapies, Stephan said it would be possible to delay the disease onset by years rather than squeezing out a few months of benefit with drugs like Aricept. If the average delay of onset were five years, the prevalence of the disease could be cut in half, saving a staggering $1 trillion per year. The *APOE* gene, included in the Navigenics panel, provides about 65 percent accuracy in predicting Alzheimer's, which should increase as other related genes are discovered.

The effectiveness of emerging therapies and lifestyle measures in warding off Alzheimer's is controversial, but Agus wasn't about to apologize for including Alzheimer's as one of Navigenics's "actionable" conditions. "Listen," he said. "My grandmother and grandfather had Alzheimer's. I want to put interventions in if they're low risk and have potential. Maybe they won't work, but I want to make that decision." In prostate cancer, Agus doesn't pretend he knows if the screening/surgery/radiation regimen always saves lives, but he's seen cases where it does. "We need the paternalism to go away," he said. "Maybe it means that for these unproven things, we pay out-of-pocket for it. Insurance companies in the long run want to pay for prevention. The problem is we all switch insurance companies all the time. If I'm an insurance company, I'm saying, why should I pay?"

"Of course, no payers in the U.S. have jumped on that bandwagon," agreed Stephan. "We've talked to the 'Blues'—they understand prevention and how it will help them, but then they say, 'If we're going to pay for this for all our customers, we're going to subsidize the whole system and we're going to go broke.' " Stephan suspects there will be more amenable partners in the state-sponsored, single-payer systems such as those in the United Kingdom and Canada. If Navigenics can show that personal genomics works, then, he argues, there's a chance it can be synchronized across all the health insurers in the United States.

After the first twelve months, Agus admitted that the Navigenics business model had morphed. Agus had assumed that "if you launch it, they will come. I thought everybody would jump on board just like cholesterol testing." The number of clients was not disclosed, but even $999 was a lot of money just to receive probabilistic information on less than twenty diseases, particularly when similar information could be obtained elsewhere for half the price. There was also less use of the professional counseling service than he'd expected, though he believed it was a key feature to help interpret the information. By analogy, Agus said he felt naked in the early days of the Internet, but "AOL made me feel clothed. In the end, you need a genetic counselor or a doctor to feel clothed." He'd also assumed the biggest demographic would be twenty- and thirty-year-olds. "I remember being one of them—I didn't have a doctor, I wanted to be involved in my health care." Actually it was older adults who used the tests the most.

Nevertheless, Navigenics was making inroads in working with medical organizations and corporations, and some corporations, including Microsoft and Life Technologies, were offering subsidized Navigenics service to executives. "If they can keep their employees healthy until they retire, they've done well. Then think of the data coming back. . . . I can go to Microsoft and say, '13 percent of your employees are at high risk of breast cancer, 8 percent of diabetes.' " The corporation gets healthier staff that work longer and pays less in health care costs with fewer catastrophic events. "The penetrance rate is amazing, absolutely astonishing," said Agus, perhaps because some people prefer the comfort of using a service sanctioned and managed by their employer rather than entrusting their credit-card details to an unknown vendor. Navigenics also struck a promising partnership with MDVIP, a concierge medical service, as well as launching a twenty-year study with the Scripps Research Institute to evaluate the long-term impact of personal-genomics services on both physicians and patients. How are doctors going to react to this information? Will patients rush to get a colonoscopy and a Lipitor prescription, then forget about their results?

Stephan laid out an ambitious program for where he wants to take personal genomics. "The holy grail is to sequence the whole human genome [of a patient], put it in a big computer, push a button, and have a rank-ordered list of disease predispositions pop out the back," he said in typically measured tones. "It will supplant newborn screening and all molecular diagnostics." With this trove of information in hand, patients will be able to ask presymptomatic questions to minimize their risk of developing disease.

Stephan would like to sequence a newborn's genome and release that information when the time is right. Once Navigenics has a client's full sequence, the client can receive regular updates as the role of newly discovered gene variants in disease is validated. Stephan's bottom line is this: "At some point in your life span, everyone will get their genome sequenced and get risk mitigation. Think of it as newborn screening for adults."

"You're going to see a call for public service," said Agus. "That includes taking your genome and putting it into the public databases. The folks who are doing it now—the Watsons, the Esther Dysons—they're heroes! That's a call to service we all should answer. It's not just walking in the Muscular Dystrophy Telethon, it's giving information back, with the correct annotation, and empowering the databases."[32]

Stephan and Agus have no doubt that Navigenics will eventually offer whole-genome sequencing. But auditing those 3 billion data points will be a massive challenge. Even now, Agus said, the hardest thing Navigenics has to do is "throw out 99.9 percent of the literature," because it's better to make no turns than a wrong turn. While most human disease—up to 90 percent—is accounted for by a handful of familiar conditions such as heart disease, diabetes, obesity, and cancer, caused primarily by common DNA variants, the remaining diseases are made up of a long tail of rare mutations. Whole-genome sequencing will not only reveal the dark matter that contributes to common disease but the rare "private" mutations responsible for other disorders.

Nevertheless, Agus admitted to having two big fears about personal genomics. One is the competition: if some companies stumble, they could take the whole personal-genomics field with them.

"You only get one shot at this," he said. At the Navigenics launch party, Al Gore likened the criticism of the field to the media fear-mongering in the early days of in vitro fertilization (IVF), yet now there are 1 million to 2 million IVF procedures annually. "If somebody screws up, it could really harm the field and make it go backwards," said Agus. But Agus is generous toward at least one of his competitors: he's a genuine admirer of 23andMe. "It's the nicest thing in the world when the people in your space are honest and good and care about doing good and helping the world. I couldn't have a better ally in the space." Avey and Wojcicki were "two of the best people I know," he said.

Agus's other worry was the failing economy, which could stall any start-up, and in November 2009 dragged deCODE into bankruptcy. "I just hope in my heart we didn't do this too early," he said. "I hope Navigenics and 23andMe can stay around. The Navigenics people are fantastic. That's why we did this. We do want to change the world."

CHAPTER 4

DNA Dreams

The competition to make the $1,000 genome a reality was kicked off in earnest by none other than Craig Venter six months before the National Institutes of Health (NIH) declared "mission accomplished" on the Human Genome Project (HGP). On October 2, 2002, Venter hosted the opening session of a major genetics conference at the Hynes Convention Center in Boston, which had the provocative title "The Future of Sequencing Technology: Advancing Toward the $1,000 Genome."[1] It was the first time that the many hundreds of scientists packed into the hall had come across the term "the $1,000 genome," although it had actually been coined the year before. The term was used at an NIH retreat held in December 2001 for a session organized by Eric Green, and had been discussed by other institute officials, including Mark Guyer and Jeffrey Schloss, a few months beforehand.[2] In the UK, former Solexa CEO Nick McCooke recalls that he started talking about the $1,000 genome in 2000 and used the phrase in fund-raising proposals in 2001—although he stops short of claiming he first used it.[3]

In his introduction, Venter cannily defined the next big question for genomics: Who would end the fifteen-year monopoly in the sequencing market that had been enjoyed by Applied Biosystems? Who could slash the cost of sequencing a human genome a thousandfold—from tens of millions of dollars to less than $100,000? And who dared to dream of a $1,000 genome?

Like an episode of *America's Got Talent*, Venter had invited half a dozen aspiring entrepreneurs to impress the audience with their new technologies. Five of the six guests were convinced that dramatic advances in a number of areas, including chemistry, optics, and computing, would be harnessed into an affordable bench-top instrument for ultrarapid DNA sequencing in the near future. Harvard Medical School's George Church discussed his laboratory's homegrown assembly of inexpensive, off-the-shelf technology, and he predicted they'd be able to offer a full genome scan for a remarkably precise $710, including $150 for electricity, within ten years. Susan Hardin, founder and CEO of VisiGen Biotechnologies in Houston, said her goal was to sequence a human genome in a couple of hours for under $1,000—how about $995? Her first instrument was just a few years away, she said, and would offer real-time sequencing of single DNA molecules. Representing the old guard was Trevor Hawkins of Amersham Biosciences (now GE Healthcare), who could see existing Sanger technologies delivering a $30,000 genome within a few years, "but to get us to the next level will take a new technology."

In fact, scientists had dreamed of inventing a breakthrough approach to DNA sequencing that would crack the $1,000 genome barrier long before Venter's symposium or the completion of the Human Genome Project (HGP). In 1987, Nobel laureate Walter Gilbert announced the launch of Genome Corp., the idea being to build a sequencing factory in Southeast Asia that (shades of Celera a decade later) would license sequence information to pharma companies. Even though Gilbert's dreams crashed with the stock market, he staunchly believed that the genome would be completed by the end of the millennium. In what he called "a vision of the Grail," Gilbert prophesied that the sequence of an individual's genome would soon be squeezed onto a single compact disc. "One will be able to pull a CD out of one's pocket and say, 'Here is a human being; it's me!' " Gilbert said this knowable collection of information would forever change our view of ourselves.[4]

The same year that Gilbert launched Genome Corp., an MIT-trained engineer named Kevin Ulmer launched a tiny start-up, simply called SEQ (pronounced "seek"), based in the unheated basement of his house overlooking Cohasset Harbor on Boston's south shore. Ulmer, an expert on sequencing in his own right, was convinced he could devise a better way, contemplating ideas—visualizing DNA with heavy uranium atoms using electron microscopy or threading DNA through submicroscopic nanopores—that are still being pursued a quarter of a century later. One thing was certain: DNA sequencing was an information-service business. "The very first business plan for SEQ talks about doing genomes at a price your medical insurance would reimburse," he recalls. "I was talking about the $1,000 genome in '87."[5]

Ulmer settled on a method of sequencing single molecules of DNA, using an enzyme called exonuclease to nibble off bases from an unfurled DNA strand like a PacMan video game. If the newly released nucleotides could be trapped and identified, SEQ would have the makings of the first "next-generation" sequencing system. But after setting up SEQ's first laboratories and hiring some leading scientists from AT&T's Bell Labs, Ulmer resigned, frustrated that SEQ had been unable to catch the investment wave of high-profile genomics start-ups such as Human Genome Sciences and Millennium Pharmaceuticals. "We were sort of three-day-old fish, shopworn, and the technology was too complicated," Ulmer admits.

Just as Ulmer was abandoning his attempts at single-molecule sequencing, a young medical student named Eugene Chan had become obsessed with a similar notion. Chan was not merely optimistic about slashing the cost of DNA sequencing, he all but guaranteed it. At Harvard, Chan, the son of a pharmaceutical chemist, got through organic chemistry classes taught by some of the world's leading minds thinking, "Dude, this is really easy!" By the time he crossed the Charles River in 1996 to enroll at Harvard Medical

School, he was fascinated about the prospect of sequencing a person's genome. He explained, "There was no—even to this day—problem more exciting than the $1,000 genome in science, actually. Pick up *Science* magazine or *Nature*, you flick through it and ask: What is the one transformational discovery that will change medicine? It's the $1,000 genome!"[6] But Chan was alarmed at the pedestrian pace of the HGP and felt there had to be a way to sequence faster and cheaper. He devoted one semester to his medical studies before his sequencing obsession overwhelmed him. Quitting his classes, he barricaded himself in his tiny dorm room, and buried himself in books and journals, trading physiology for physics and neurology for nanotechnology.

Chan's goal was to divine a way to untangle single DNA molecules—a feat far less trivial than it sounds. In the cell nucleus, DNA molecules naturally exist as chaotic, tangled balls of nucleic wool, like a tangled extension cord, seemingly with no beginning or end. "The best way to think about it," said Chan, "is a slinky that's really long, floating in solution. It kind of wobbles and goes back and forth. It's a much more complicated object than a ball of wool or spaghetti."

For months, Chan became an obsessed recluse, and he admitted his behavior was "pretty wild," but he finally emerged with a two-hundred-page scientific plan that he was convinced was going to work. By stretching out the newly straightened DNA molecules, tagging them with chemicals, and then running them past a laser, Chan believed he could read out the DNA sequence in record time. His confidence was backed by MIT professors, including Robert Langer and Alex Rich, experts in medical engineering and DNA biophysics, respectively, who grilled the twenty-three-year-old student as if he were defending his thesis. "If [Chan] could do the sequencing that rapidly, that would be a change-the-world kind of thing," Langer told *Scientific American*.[7]

Chan quit medical school and founded U.S. Genomics, and Langer and Rich joined his scientific advisory board. Renting incubator space at Boston University, Chan took only six months to produce linearized DNA molecules, just as he had imagined. In a feat of nanotechnological engineering, Chan passed the DNA through a series of microscopic pillars that snagged and unfurled the DNA molecules before funneling them into a narrow "interrogation channel" where they zipped under the eye of a laser at about 1–2 cm/second.

An intense media spotlight suddenly shone on Chan. There was a profile in *Scientific American*, an interview in *Newsweek*,[8] and selection as one of 2002's "Best and the Brightest" by *Esquire* magazine, all fascinated by the handsome scientific rock star speed reading DNA.[9] "It took us from the dawn of time to the year 2000 to map the human genome," the *Esquire* profile began. "Eugene Chan . . . wants to do it while-u-wait." The following year, *Technology Review* anointed him one of the top 100 young entrepreneurs,[10] and investors merrily pumped another $25 million into his company.

Around that time, I visited Chan at his headquarters in a nondescript office park north of Boston. We walked into a small room to observe a test run of the "Gene Engine," an imposing machine the size of a Sub-Zero refrigerator. A technician loaded in a DNA sample, punched a few buttons, and within a few seconds, images of tangled white balls of DNA were rushing by on the machine's monitor, almost like magic emerging as neat straight rods running through the detection chamber like a subway train. It was a stunning demonstration.

As Chan described it at Venter's 2002 symposium, "The DNA is uncoiled, unfurled, and read past a fixed reader, just like a movie reel spools over a projector," scanning DNA strands up to 200,000 bases long in less than a second. Chan said he would slash the cost of genome sequencing. "The molecules move by at 30 million bases of DNA per minute," he said. "In a forty-minute time frame, you get about 3 billion base pairs of DNA going through the system." Three billion bases, he didn't need to say, was the size of a human genome. Moreover, with further enhancements, Chan reckoned he could slash the time for scanning to five minutes. "Our goal is to be able to read your genome instantaneously," said Chan.[11]

His self-assurance suggested that doing so was just a few years away, a mere formality. About the only person in the audience not buying the story was one of Chan's colleagues, Mike Shia, who was dismayed that his boss could be making such brazen claims publicly with no sequence data to back them up. "Eugene wanted to do the whole human genome, but he never did any real science," he said. "He kept calling it a *sequencing* technology. It was never sequencing technology."[12] In fact, getting the strands straightened out and scanned quickly was only one step toward faster sequencing; it didn't solve the problem of actually reading the sequences faster. And without that, Chan's Gene Engine had much more limited appeal than all the hype suggested. The U.S. Genomics advisory board told Chan that his platform was too futuristic. By March 2003, a seasoned biotech executive, Steven DeFalco, had taken over as the CEO of U.S. Genomics and abandoned the goal of becoming the provider of the "$1,000 genome," switching instead to applications for the biodefense industry. Chan was left with no direct responsibilities; he resigned from the company and completed his MD and an internship at Brigham & Women's Hospital. He is still contemplating ways to integrate genomics and medicine.

The Wikipedia entry for Jonathan M. Rothberg, the founder of 454 Life Sciences, once credited him as "the first person to sequence an individual human genome" and thereby "initiating the age of personal genomics and individual genome sequencing."[13] It has been tempered considerably since then, but probably not by choice. As I sit in Rothberg's stunning art deco office in his own institute devoted to genetic diseases, I'm reminded less of a captain of industry or biotech impresario than Tom Hanks in the movie *Big*. Everything about Rothberg, from his unkempt appearance to his office decor to his extraordinary indulgences, suggests a boy trapped in a man's body. His black, shoulder-length hair doesn't appear to have been combed in years, and his pink designer shirt is at odds with his solar system tie, white Mickey Mouse socks, and hiking shoes.[14]

His office looks like the set of *Happy Days*—his desk is a diner bar littered with executive toys. Along one wall is a scarlet leather sofa from a 1959 Cadillac, while along another are wine bottles from the nearby vineyard Rothberg bought from the former CEO of Tiffany & Co. I met with Rothberg at his nonprofit foundation, the Rothberg Institute for Childhood Diseases, in Guilford, Connecticut, which supports research into orphan diseases, principally tuberous sclerosis, a hereditary disease that affects his oldest child, Jordana.

Rothberg proudly opens a photo album and shows me a photograph taken a few months earlier of Watson standing in front of a DNA sequencer made by 454, which had read Watson's DNA. Rothberg had said publicly at the Baylor press conference that the retail price of Project Jim would be about $1 million; in a speech a few weeks earlier, he had said the internal cost for Watson's genome was about $200,000.[15] "This was really good for me because basically it was closure. I had started 454 to sequence an individual," he said. But the same week that Watson's results were released, the diagnostics company Roche acquired 454 for $150 million, leaving Rothberg free to pursue other ambitious ideas.

Jonathan Rothberg was born in New Haven, Connecticut, in 1963, one of seven children, destined to be a chemical engineer like his father, Henry Rothberg. In the 1950s, Henry had developed a revolutionary adhesive for stone installation, and founded Laticrete International. Today, Laticrete's adhesives are used in major construction projects worldwide. The company is now run by Jonathan's youngest brother, David.

Rothberg had studied chemical engineering at Carnegie Mellon University in Pittsburgh, where he did his first DNA sequencing experiments in 1985. He was torn then about whether to continue to work on sequencing or switch to cognitive psychology. The big question to him was whether biosciences would make more progress over the next decade than the study of the mind. Rothberg chose to pursue a PhD in biology at Yale, where he joined the lab of fruit fly geneticist Spyros Artavanis-Tsakonas. Rothberg had the first personal computer in the lab and even drove a Maserati, a gift from his mother, although Artavanis-Tsakonas roared with laughter as he recalled it was an automatic shift, which rather defeated the purpose.

In the early 1990s, there was still a pervasive feeling among most scientists that abandoning academia for biotech was a bit tacky. But although he'd published some good papers, Rothberg was always thinking about business and woke up his boss to the existence of biotech. Artavanis-Tsakonas launched a San Francisco biotech company called Exelixis—one of the first genomics start-ups—and invited Rothberg, still a graduate student, to join as the chief science officer. But Rothberg declined, intent on starting his own company instead.

In 1991 he set up his company, CuraGen, in his basement with the goal of using information from the fledgling Human Genome Project (HGP) to make better drugs. "He'd love to be seen as a guy who came from nothing and built something, when in fact his parents were major shareholders from the start," said Martin Leach, a British computer scientist at Merck who worked at CuraGen for nine years.[16] Also investing was Rothberg's brother Michael, who gained a 25 percent stake in the company, and his college friend Greg Went. From modest beginnings, Rothberg racked up a number of impressive accomplishments. He invented a patented method for studying gene expression by looking up activity patterns in a database, which he dubbed "GeneCalling." And then a number of major investments in the company quickly followed: $2 million from Genentech,[17] $10 million from George Soros's Quantum fund, and $15 million from Biogen. By 1998, Rothberg had raised some $600 million and estimated that his three major investors (Biogen, Pioneer, and Genentech) each made $100 million.

Rothberg had brought the same systematic approach to his personal life as to his business. After deciding he had to marry an MD, he held a huge party to meet the Yale MDs. Among the guests was a Canadian physician named Bonnie Gould. The couple married in 1995. In 1997, Rothberg's family was devastated when his first child, Jordana, was diagnosed with tuberous sclerosis. "Jonathan is such a perfectionist that it's a flaw to have something like this," Laura Davis, Rothberg's personal assistant, suggested. "He could not admit to any personal weakness."[18] Rothberg's response was typically audacious: with support from his family, he founded the Rothberg Institute. One of its first projects was to set up a distributed computing network (courtesy of one of his nephews) to tap into unused processing power from the public's home computers to identify potential drugs.

As I admired the framed patents and magazine covers hanging in Rothberg's office, my attention was especially held by one photograph. At first glance it looked like an aerial photograph of Stonehenge, except that this monument is just a couple of miles away. "I built that for my kids," Rothberg shrugs, as if it was entirely routine to import 700 tons of Norwegian granite and sculpt, polish, and arrange it according to celestial factors on one's own oceanfront property. The sheer existence of the *Circle of Life* spoke volumes about his passion, determination, and exuberance.

When Rothberg and his wife, Bonnie, began redeveloping an eleven-acre former quarry overlooking Long Island Sound in 2003, Bonnie had visions of building a soccer field. More ambitiously, Rothberg wanted to build a small astronomical observatory. But the local zoning board rejected his plans, designed by Cesar Pelli, judging not unreasonably that the proposed 35-foot-high structure might not blend in with its surroundings. Not one to take rejection lying down, Rothberg devised a new plan. If he put up a work of art instead, he wouldn't need planning permission. "It might be that you couldn't build Stonehenge because you can't get the stones," Rothberg said. "It might be that the technologies are lost. Or it might be that you'd get lost in legal hassles. Anyway, I wanted to know."[19] He hired sculptor Darrell Petit to create a sophisticated astrological clock based on "certain astro-archeological concepts of Stonehenge."[20] The *Circle of Life* was like "a complex watch . . . that exhibits beauty and functionality and demonstrates that these structures can function as powerful astrological instruments with substantial predictive abilities," Petit told *Stone World*.[21]

After the shipment of 700 tons of blue pearl Fjord Norwegian granite across the Atlantic, Anthony Aveni, a Colgate University astronomer, arranged the stones based on the Rothberg children's birthdays and other astronomical events. The *Circle of Life*, completed in July 2004, consisted of twenty upright pieces, 4 meters high and weighing 22.5 tons apiece, and twelve lintels. Petit said the structure would endure 10,000 years because it had "the energy of millions of years that would reflect the sun and stars of the expansive universe."

As CuraGen filed its initial public offering (IPO) in 1998, friction began to surface within the CuraGen Board. Rothberg was a much better inventor and creator than day-to-day manager, and the board began casting for a new CEO. Suddenly, the search was canceled. The markets exploded as Craig Venter and Celera led the genomic gold rush. "People like [Incyte's] Randy Scott, [Human Genome Sciences'] Bill Haseltine, Venter, and I become the rock star of genomics. It was a rocket ship," Rothberg said. "In 1999, we were the number one stock on the NASDAQ." At its peak, CuraGen was worth $5 billion, more than American Airlines.

But another event in 1999 quickly brought Rothberg down to earth. In July his first son, Noah, was a "blue baby" and rushed to newborn intensive care. That night, Rothberg nervously paced and fidgeted in the waiting room. "Why can't we just sequence his genome and know everything is fine or not?" he wondered to himself. DNA sequencing still relied on Fred Sanger's twenty-year-old method, and since the launch of the HGP in 1990, only a small percentage of the genome had been sequenced.

Rothberg then reached into his briefcase and pulled out a copy of *InfoWorld*, a computer magazine. The cover had a picture of the latest Intel microprocessor boasting 44 million transistors, and his mind began dancing with possibilities. Scientists had invented the transistor because vacuum tubes burned out too fast. And although transistors lacked the power of vacuum tubes they were cheaper and better. "Everyone said that transistors were inferior," he said. "But transistors would move to a miniaturized, integrated platform." For years researchers had been trying to find a way to make Sanger sequencing work on computer chips, and failed. Rothberg concluded: "Forget it—I'm going to move an inferior technology to a chip." He would make DNA sequencing the equivalent of the integrated circuit.

Noah made a swift recovery, and Rothberg was able to spend his paternity leave developing his idea. If Rothberg invented the sequencing equivalent of the integrated circuit, then Matthias Uhlen should get credit for inventing the transistor. Uhlen is one of the developers of pyrosequencing, the light-based detection method Rothberg selected as the only commercially available alternative to Sanger around 2000. Pal Nyren had conceived the process while he was a postdoc in Cambridge, England. One rainy January evening in 1986, Nyren was cycling home when the idea for an alternative DNA sequencing method flashed into his mind. His wife listened to his brilliant idea, all the while thinking he looked like the Disney character Gyro Gearloose's little assistant with a light bulb for a head. Nyren returned to Stockholm after his postdoc, but it took a decade more to fully develop the method.[22]

The idea was to focus on a small molecule by-product, pyrophosphate, that was produced every time a new base was incorporated into a DNA chain. The Swedes would add one base of DNA at a time (first A then C then G then T) and, starting with the release of pyrophosphate, use a clever chemical domino effect involving an enzyme called luciferase, extracted from fireflies, to turn the chemical cue into a pulse of light. The amount of pyrophosphate, and thus the amount of light, released is directly proportional to the number of bases incorporated. Nyren, Uhlen, and other key scientists founded a company to commercialize the method, now known as pyrosequencing AB.

Although Rothberg saw limitations in the method, miniaturizing the process nevertheless brought two big advantages: he would save on reagents, and the process removed the limitation that stopped the technique from full-blown sequencing. Rather than use enzymes (which interfered with the sequencing) to remove by-products before each new cycle, Rothberg would simply wash the by-products away. "Those guys missed it," he said. "My idea was to move pyrosequencing to a chip, taking advantage of Moore's law. And this was completely novel at the time—I had to keep up with it. We couldn't do individual sample preps or the costs for the genome would be $100 million to prep it and $1 million to sequence it."

Looking back, Rothberg says, "I made two contributions at that moment that are now ubiquitous on next-gen [sequencing]. First, I knew you had to get rid of bacterial cloning. The second was sequencing in the miniature." Traditional sequencing methods relied on using bacteria to purify and grow DNA fragments. But Rothberg favored a variation of the polymerase chain reaction (PCR) for amplifying DNA called emulsion PCR. In this method, he separated out individual DNA molecules by shaking them in a suspension of water and oil—like the opposite of mayonnaise (oil drops in water)—so that each DNA fragment became embedded in a separate water droplet, like a miniature test tube. Each DNA fragment was then amplified so that each water drop contained some 10 million identical copies of the DNA fragment suitable for sequencing. Rothberg would then drop into the mixture tiny beads that became enveloped with the DNA. These beads, so small that four would fit on the end of a single human hair, were then spread out into thousands of tiny wells, allowing Rothberg to run hundreds of thousands of sequencing reactions in parallel, something that had never been done before.

Rothberg decided that 454 Life Sciences would not be monopolized by CuraGen. "454 was my playground, and I had a strategy," he said, and that included getting the company separate from CuraGen. Rothberg kept the origin of the "454" name a fiercely guarded secret. "It's my code in case somebody stole it and they claim that they created it," he said. "I know that sounds crazy!" Some CuraGen staff speculated the name was a riff on *Fahrenheit 451,* while Jim Golden, 454's bioinformatics director, said the name derived from a financial spreadsheet: the total revenues projected for the first five years fell in row 454. Martin Leach offered a more romantic possibility: in 1999, while Rothberg was writing patents, Cisco Systems had acquired Cerent, the producer of an optical transport device called the Cerent 454 that allowed Internet service providers to boost bandwidth delivery and revolutionize Internet traffic. Rothberg wanted 454 to be the Cisco of biology and provide the necessary bandwidth for high-throughput genome sequencing. (Rothberg is still mum on the subject.)

The launch of 454 was conducted under similar secrecy—even Davis didn't know what all the fuss was about. When candidates came to interview, about all her boss would tell them was, "I have an idea for a technology. It's going to be as big as MRI someday." It says something for his charisma and powers of persuasion that the people he wanted usually signed on.

One of Rothberg's brilliant recruits was Scott Helgesen, an immensely talented software engineer who had just seen a multimillion dollar offer for his Internet start-up vanish in the dot-com crash. As Helgesen tells the story, he got the job at 454 due to a curious entry on his résumé. "I got my job at 454 because of my effects work on *Men in Black*," Helgesen said.[23] Indeed, during his interview at 454 headquarters, it was the only project anyone wanted to hear about.

In 1997, Helgesen had formed a company to develop a technology that allowed people to turn amateur photographs into professionally edited films simply by answering a set of questions. To finance the project, Helgesen and his friend Brad Carvey (the brother of *Saturday Night Live* star Dana Carvey and the inspiration for Dana's character Garth in *Wayne's World*) took a flier on making movie special effects, and they were commissioned to create the opening computer graphics dragonfly sequence for *Men in Black*. Helgesen could sense during his interview that he had the job once the topic came up. "It had nothing to do with my knowledge of genomes or sequencing. All the questions had to do with *Men in Black*." The only downside was leaving his home in sunny New Mexico for the cloudier climes of Connecticut.

Helgesen's task was to build a software group from scratch and design the critical 454 sequencer software to process the photographic images produced from the luminescing beads into sequence data and perform the final genome assembly. He brought in cutting-edge technology using chips called field-programmable gate arrays (FPGAs), which no one at 454 even pretended to understand, and called his plan *reconfigurable computing*. He changed the digital logic inside the computer hardware thousands of times per sequencing run to do all the image processing in real time instead of relying on conventional software. Helgesen saw a way to reduce the processing demands to be able to process data on the instrument instead of requiring a huge computer farm. "It's the only way it [sequencing] will show up in a doctor's office," he said. "The doctor can't be hooked up to some big mega server farm somewhere."

Meanwhile, 454 scientists and engineers were struggling with the chemistry. At first, 454 could barely generate any light out of the wells during the sequencing reactions. "It was like, here's an image we think we got some light out of!" said Helgesen. Nobody knew exactly what was in a particular well. Was only one bead sitting in each well? Were the wells sufficiently cleaned between cycles? Was there contamination between the wells? The CEO, Richard Begley, implored his colleagues to imagine themselves inside the well and think about what was happening. "Be the well," he chanted. "Be the well." Christopher McLeod, who would later succeed Begley, remembered sequencing a run of five bases as a major landmark. Eventually the read lengths grew to twelve bases, then to twenty-five and 454 was finally on its way.

In 2003, 454 tackled its first genome sequence: the 30,000 bases of a tiny adenovirus. It took less than a day. In a triumphant press release, 454 crowed that its success "marks the first time that a new method has been used to sequence a whole genome since

Walter Gilbert and Frederick Sanger won the Nobel Prize in 1980 for the invention of DNA sequencing." A story in the *New York Times* quoted the Department of Energy's Eddy Rubin as saying the method "would never scale." But Baylor's Richard Gibbs said, "This is going to be big." Rothberg naturally elected to collaborate with Gibbs going forward.[24] Next, 454 turned to sequencing a bacterium.

Critics charged that 454's technology was naturally inferior to the Sanger method—shorter read lengths, poorer accuracy and genome coverage. But Eric Lander, director of the country's largest genome center at the Broad Institute, was excited by 454's technology, and in 2004, he and Rothberg teamed up to look at a larger fungus genome. Meanwhile, 454 had started a service business that offered sequencing to customers such as Johnson & Johnson. They reported success sequencing both bacteria that were resistant and were not resistant to a new drug for tuberculosis, revealing a mutation in the drug target.[25]

In August 2005, Rothberg and colleagues achieved a highlight in any scientist's career when they published the blueprint of 454's technology in *Nature*.[26] First among the fifty-six coauthors (including Rothberg, Begley, and Helgesen) was Marcel Margulies, 454's engineering chief, whose previous job was working on the Hubble space telescope. The paper showed that 454 had assembled a bacterial genome sequence more than fifty times over as a proof of concept, with an instrument that was able to generate 25 million bases in just four hours. The journal editors concluded that the 454 instrument signaled a new era in genomics—if not all of biomedical research. The Genome Sequencer 20 (GS20), so named because it sequenced 20 million bases per run, contained 400,000 wells, each of which could sequence a piece of DNA up to 100 bases in length.

On the eve of the publication, I reached Rothberg by phone while he was vacationing with his family in Montreal. To say he was excited was an understatement. This was arguably the pinnacle of his career. Over five years, 454 had spent $50 million and had hired 100 talented scientists, engineers, and programmers in pursuit of a vision. Not only had they developed the first commercially available next-gen sequencer, but they had sold these machines to many of the most prestigious genome centers in the world, including the Broad Institute, the Wellcome Trust Sanger Institute, and the J. Craig Venter Institute. 454 had delivered its first GS20 to the Broad in early 2005. "I still have the check, like a good retailer," said McLeod. 454 was selling its machines as fast as it could build them, despite a list price of $500,000.

To Rothberg, this was a defining moment in the history of medical science: "It's completely analogous to personal computers displacing mainframes. Now, anyone can have their own genome center. If you can miniaturize something, then everything gets cheaper and faster. Since Fred Sanger won the Nobel Prize in 1980, it's been a race to find a new way to sequence DNA cheaper and faster. There's never been any technology to pull that off."

Rothberg was giddy with potential applications ranging from bioterrorism, to third world drugs, to sequencing HIV in AIDS patients, to gauge resistance. "There's no more waiting in lines at genome centers. Anything you want, you can sequence pretty instantly," he said. To mark the publication of the *Nature* article, I posted an online story with the unscientific title, "Fantastic 454." Less than ten minutes after I'd e-mailed Rothberg the link, he replied on his BlackBerry. "Kevin," Rothberg thumbed, "I love your title. My kids also love it." He went on to reiterate the advantages of 454's method against its rivals, especially the latest paper from George Church's group at Harvard, published that same week in *Science*.[27] A single 454 machine could match 100 Sanger instruments. "Anyone buying one of our machines has the throughput of a major genome center," he wrote. And he closed in characteristically ebullient fashion:

> Our technology is like the personal computer, democratizing sequencing. It is also based on a chip that we manufacture, and each year we will be able to do longer sequencing reads in each of the separate reactors, and more reactors per chip—just like

> Gordon Moore said would happen for computer chips. [We] expect that our technology will continue to advance to the point that it will be useful to have your own genome sequence while you wait at the doctor's office.

That said, Rothberg admitted 454 could do much better. It needed longer sequences from each well, up to 200 or 400 bases; more wells per square millimeter, raising the total throughput per run from 20 million bases to 100 million and beyond; and improved sequence accuracy, which stood at a very unsatisfactory 96 percent. But for now, Rothberg felt vindicated and victorious. He signed off with a flourish. "We won the race. Everyone may not be happy with that, but we are."

Rothberg had just one nagging concern: the fear that a mystery person was hatching a new technology in his or her basement, just as Rothberg had done fifteen years earlier. Scott Helgesen was even more worried. He thought 454 was being complacent and should be following the advice of Netscape founder Jim Clark, who believed that a company creating a disruptive technology should be comfortable with creating a new technology that would figuratively put it out of business—because that was what the competition was doing. In January 2006, Helgesen e-mailed 454 management a 17-point memo listing the critical next steps in 454's evolution: more research on new computer chips, a Web interface and remote cell phone monitoring, an open-source movement to share source code, the development of a smaller, more affordable machine, and software development for single-molecule sequencing. "We need to start getting paranoid about potential competitors that are nipping at our heels," Helgesen warned. "We need to create the NEXT BIG THING!"[28] Two weeks later, he was out of a job, dismissed on a long-distance teleconference.

For Helgesen, the 454 sequencer was like the Apple Newton, a wonderful innovation that was quickly overtaken by better devices. "We created the next-gen cool technology and it worked on all fronts," he told me, "and then we sat back and watched others leapfrog past because our executive managers were not smart enough to understand what it takes to stay out front." Helgesen fired off one final cathartic e-mail reminding Rothberg and McLeod of his team's accomplishments—the instrument control, the signal processing and sequence assembly software, even the design of the marketing materials and packaging. There was no reply.

Rothberg had three major ambitions when he founded 454. The first was the ability to sequence individual genomes, prenatally or at birth, so that parents could know their child's DNA. Another was the ability to decode the genomes of new organisms, which would showcase 454's greatest competitive advantage: its ability to sequence individual DNA strands hundreds of bases in length, while competitors were stuck on less than fifty bases. (When 454 introduced its new-and-improved FLX instrument, the marketing team printed T-shirts with the slogan, "Length Matters.") Rothberg's other abiding interest came into focus only after the terrorist attacks of September 11, 2001: Metagenomics is the genetic study of entire ecosystems of organisms, such as the population of microbes in the human gut or viruses infecting an insect. He wanted to sample the air and catalog everything that was in "the genosphere."

A quite unexpected application turned not so much on the future of genome sequencing but on the decoding of ancient genomes—with profound implications for the study of human biology and evolution. In 1996, geneticists from the Max Planck Institute in Leipzig, Germany, led by Svante Pääbo, published the first detailed analysis of a wisp of Neanderthal DNA, painstakingly purified and analyzed from the original Neanderthal skeleton discovered in the mid-nineteenth century in the Neander Valley. The study appeared in an issue of the journal *Cell* that sported a Neanderthal skull on the cover and the provocative headline, "Neanderthals were not our ancestors."[29] Pääbo's team had coaxed sufficient traces of DNA from the fossilized bone to compare with the corresponding mitochondrial DNA traces in humans, and found a large number of sequence variations compared with modern humans, knocking the theory that humans and Neanderthals might have interbred during their coexistence across Europe before the demise of the Neanderthals some 30,000 years ago. But why stop there? "The dream of Neanderthal biology isn't just to sequence Neanderthal," explained Rothberg. "It's to see the genes that are different between humans and Neanderthal," in search of genetic clues for the origin of speech and other quintessential human traits.

454's entry into the ancient-DNA field was sealed during a phone call with Pääbo while Rothberg was strolling, appropriately enough, around his *Circle of Life*. "Let's sequence a dinosaur!" Rothberg pleaded, saying he had wanted to clone a dinosaur even before *Jurassic Park*. "You're crazy," Pääbo replied. Rothberg changed tack. "Okay, send me Neanderthal," he said. Pääbo wouldn't even consider trusting this eccentric whiz kid with his precious Neanderthal material, explaining that contamination was almost impossible to contain. As a compromise, however, he agreed to send DNA samples from a pair of 40,000-year-old extinct cave bears.

Pääbo had nearly given up his dream of sequencing *Homo neanderthalis,* but 454's success sequencing the cave bear DNA convinced him otherwise. The Neanderthal ancient-DNA fragments were essentially single molecules, less than 120 bases long, ideally suited to the 454 process. "It was a match made in heaven. That's why he didn't hang up on me!" said Rothberg. In February 2009, Pääbo presented preliminary results of the Neanderthal genome based on fossils from Croatia.[30] Incredibly, analysis of a much-studied speech gene called *FOXP2* showed that the Neanderthal version was identical to the human gene. "There is no reason to believe they couldn't speak like us," claimed Pääbo.

In May 2010, Pääbo and colleagues published the first draft of the Neanderthal genome, based on assembling 4 billion bases from less than half a gram of bone powder from three female Neanderthal bones excavated from a cave in Croatia where they had lain undisturbed for 38,000 years.[31] A major goal is to scour the two genomes to pinpoint when DNA changes swept through the human population in the hundreds of thousands of years since the two lineages parted. Alas, the Neanderthal genome has yet to reveal any clues as to what caused the demise of our closest relatives.

Another fan of 454's technology is Penn State professor Stephan Schuster, a leader in the field some call *museomics*—the study of the ancient DNA of extinct species such as the woolly mammoth and the Tasmanian tiger. To study the woolly mammoth, Schuster became a bit of a hair fetishist after finding that hair (plentiful in natural history museums) is an excellent source of mitochondrial DNA. A more recent project involves a species suddenly facing the threat with extinction: the Tasmanian devil, which since 1996 has fallen prey to the rampant spread of a deadly infectious oral cancer.[32] Because of low genetic diversity, numbers of the marsupial family have declined 60 percent, and experts worry the species could be extinct by 2030. One suggested remedy is to build a fence across the island of Tasmania, much like Australia's "rabbit-proof fence" of the early 1900s, to save the uninfected devils. Meanwhile, Schuster and his Tasmanian colleagues hope that, by sequencing, they can identify genetically diverse animals and thus guide a new breeding program to maximize the fitness of the endangered population.

The Tasmanian devil isn't the only endangered critter that 454 has helped. Rothberg was particularly proud about a metagenomics analysis of disappearing honey bees.[33] Colony collapse disorder (CCD) was first reported in 2006, by an apiarist who found hundreds of hives in Florida full of honey and larvae, but deserted. The losses topped 10 billion bees and millions of hives, and threatened an agricultural catastrophe, with theories surrounding the disappearance ranging from drought and disease to cell phone radiation and pesticides. 454 and researchers at Columbia University, led by Ian Lipkin, began to search for the putative culprit(s). The study involved grinding up bees from healthy and collapsed colonies, and performing 454 sequencing on the sum DNA, which included the bee genomes and those of any microbes or parasites attacking the insects. "It's like if we took you, your shoes, your socks, your sweater, and sequenced everything," said Lipkin. "Then we need to figure out what was shoes, what was socks, and what was you."[34] The Columbia team found traces of more than a dozen viruses, fungi, and bacteria, including the Israeli Acute Paralysis Virus (IAPV), which suspiciously lurked only in the CCD hives[35]—suggestive evidence that IAPV contributed in some way to CCD.

In another headline-grabbing collaboration with Lipkin, 454 participated in a *CSI*-worthy episode to solve a bizarre transplant tragedy in Australia. A few months after vacationing in the former Yugoslavia, a fifty-seven-year-old Australian had a stroke and died. His organs were transplanted into three middle-aged women, who all died a week later. Lipkin's group sequenced samples from the deceased women's organs and, after a database search, found genetic traces of an arenavirus that is found in rodent urine. It seems likely that a normally harmless virus turned deadly in the trio of transplant patients, who were all taking immunosuppressants. The story, published in the *New England Journal of Medicine*,[36] drew rave reviews from vaunted virus hunters such as the NIH's Anthony Fauci, who praised the "spectacular power" of 454's new technology.[37]

"The *New England Journal* paper is probably the easiest thing we've ever done, but it has such wide implications for what we can do to save humankind, which is why we started this company," said 454's chief scientist, Michael Egholm, who assumed the chief evangelist role from Rothberg after Roche snapped up 454 in May 2007 for $140 million.[38] "There's really an amazing possibility now of preventing another HIV epidemic, another SARS, or what have you. We can basically sequence it all and find it long before it ever becomes an issue. Had we been around in the early '80s, when HIV started popping up, my personal conviction is that that epidemic wouldn't have existed. Had [the first HIV cases] shown up today, you'd have had a diagnostic right away." Sitting beside him, 454 CEO Christopher McLeod almost spat out his coffee in shock.

Egholm was proud that 454 had developed the first next-gen sequencer to reach the market, but admitted the first systems were far from perfect. He waved off the threat posed by rival next-generation-sequencing competitors, insisting that his sights were set on "delivering the death blow to Sanger sequencing." He thought about running an advertising slogan: WHAT DO YOU WANT: BASES OR BIOLOGY? "It's not the total throughput that matters, but what you do with it. I wouldn't trade my BMW for seven Hyundais," he said. To mark the launch of 454's latest upgrade, bringing the cost of a human genome close to $100,000, the imaginative Roche marketing team printed new T-shirts that read, "Length *Really* Matters."

Unfortunately, price also mattered, and for all of 454's headline news, rival platforms were proving increasingly attractive for high-throughput, cost-effective human genome sequencing.

When I visited Rothberg in 2007, he was the equivalent of a free agent and could barely bring himself to talk about the boardroom battles he had seen before and after resigning as CuraGen's CEO in 2005.[39] Rothberg said that the firm's management had subsequently "burned CuraGen to the ground," selling 454 for a measly $140 million when it was potentially worth up to $1 billion in a public offering, which in turn could have fueled CuraGen's drug pipeline. "I had a plan. I didn't want to leave CuraGen. I didn't want to leave 454. I'm not going to tell you otherwise."

His pride in 454 was still evident, but his new baby was a top-secret company called Ion Torrent Systems. "If I had a vision in 1999, Jim's genome has given me an exact roadmap of how to execute it. This has to have clinical impact. You can guess what my next venture will be: getting this into the clinic to affect people's decisions every day. So I'm still in the business. I haven't finished."

Rothberg was fascinated by the consumer-genomics market, and had met 23andMe's Anne Wojcicki earlier in the year after receiving his "favorite e-mail ever. It came from Wojcicki and read: 'Larry [Page] and Sergey [Brin] say I have to meet you.' Talk about that for your ego!" But he wasn't sure if they could sustain a business. Rothberg believed the target audience for personal genomics will be "the global upper-middle class." Americans pay on average about $6,000 per person each year for health care. Rothberg's son

Noah probably incurred $40,000 in bills the first week of his life. So if you could make a decision to save $10,000, why not? And there was another intriguing possibility. "It tells you who you can marry, and who you shouldn't marry," says Rothberg. "For sequencing, I do subscribe to the *GATTACA* vision: I'll be able to sequence someone shortly after having a meeting with them. You're going to want to do sequencing on the spot."

Before I left, Rothberg fetched a couple of bottles of his own vineyard's prize chardonnay from a small refrigerator outside his office. "Do you prefer oaked or unoaked?" he asked, before launching into a connoisseur's description of each wine's characteristics. So which version did the master vintner prefer? Rothberg simply shrugged.

"Oh, I don't drink," he said.

CHAPTER 5

The British Invasion

The University of Cambridge chemistry department doesn't immediately dazzle visitors with its aura of 300 years of research, Sir Isaac Newton, and, at last count, fifteen Nobel Prizes. On this gnarly November morning in 2008, a student sheepishly lets himself into a lecture hall, hoping the professor jabbing at a prehistoric overhead projector doesn't notice his tardiness. A caterer arrives pushing a trolley carrying a platter of limp sandwiches for a lunchtime seminar. "Ooh, don't those look nice," clucks a delusional receptionist.

I might have mistaken the trim figure approaching me for a graduate student, save for flecks of silver in his hair. Professor Shankar Balasubramanian greets me in a thick Scouse accent reminiscent of Ringo Starr. His fleece jacket sports the logo of Solexa, the company he cofounded, and comes in handy as we walk the two blocks to the local pub, The Panton Arms, which quickly fills with ravenous chemistry students. Since he's not planning on doing any experiments that afternoon, Balasubramanian graciously joins me in a pint of beer. "I don't know if you've met many chemists," he says, "but they're not known for their allergy to alcohol."[1]

As it happened, Balasubramanian had something special to celebrate. Just two weeks earlier, he was one of about 100 authors on a historic paper published in *Nature* describing the sequencing of the first African genome.[2] But the paper was also a belated tribute to a new sequencing chemistry that Balasubramanian and his colleague, David Klenerman, had conceived a decade earlier and had now had turned into a reality for one of the companies competing to be first to the $1,000 genome. "It's very exciting; we've waited for ten years to disclose something," he said, savoring his real ale. As a small British company, Balasubramanian and his colleagues had deliberately shied away from publicly saying too much too soon. Alongside their paper describing the first complete African genome, *Nature* had run other papers describing the first Asian genome and the first cancer genome, also identified using Solexa sequencing chemistry. Just six months after 454's belated publication of Jim Watson's genome, *Nature* had published three papers in the same issue, all featuring a rival chemistry and a new commercial powerhouse, San Diego–based Illumina, which had acquired Solexa two years earlier.

Balasubramanian emigrated to England from Madras with his family at the tender age of nine months, and settled in Warrington, roughly halfway between Manchester and Liverpool. As a teenager, he dreamed of being a professional soccer player, but when those dreams fell short, he settled on his safety school, which just happened to be Cambridge University. Ten years later, after a spell of research in the United States, Balasubramanian joined the faculty at Cambridge, studying the enzyme that replicates new DNA molecules. "We'd submitted a paper for publication and got reviewer comments," he recalled. "I had to do one extra experiment and I needed a laser. I found out there was a new guy in the department, a guy named David Klenerman, a laser spectroscopist. We had a cup of tea, he had a laser, he helped me address the reviewer comments, the paper got published [in March 1998], and then we started talking about things we might do together."

The two chemists decided to play around and try to watch DNA polymerase stitch nucleotides into a single molecule of DNA. Balasubramanian wanted to visualize the process in a controlled manner, like pausing a DVD before hitting Play. If he could add a base, pause and identify it, and repeat the cycle, he might be able to read a long sequence of DNA step-by-step, one base at a time. And drawing inspiration from American biotech firms like Affymax and Affymetrix, he could do thousands of reactions in parallel.

The genesis of Solexa came on a warm August evening in 1997. The two chemistry professors were having a few beers at The Panton Arms with two of their postdocs, Mark Osborne and Colin Barnes. Balasubramanian insisted this was out of necessity because, at the time, the chemistry department lacked suitable seminar rooms and their offices were too small. I asked him why they didn't go to The Eagle—the legendary Cambridge pub where in 1953, Watson and Crick exclaimed, "We've discovered the secret of life." "The Eagle is a biologists' pub, it's not a chemistry pub," he replied. As the beers flowed that summer evening, the quartet's ideas began to ferment. "I remember going home feeling pretty excited, as I often did after a discussion at The Panton Arms," Balasubramanian deadpanned. "The acid test, of course, is how you feel when you wake up and sober up." The next morning, he still thought it was a great idea.

Balasubramanian and Klenerman had come up with a new way of sequencing that they anticipated would lead to a 100,000-fold improvement in efficiency. Their idea was to read one base at a time by temporarily halting the growing DNA strand to read the identity of the newly incorporated base, then unblock the base before repeating the cycle. It was the makings of a digital sequencing process: a T or not a T? That was the question, repeated twenty-five or thirty times, in tens of thousands of parallel reactions.[3] They pitched the idea to the venture capital firm Abingworth using a grand total of five acetate sheets for an overhead projector. On one slide, Balasubramanian sketched a bunch of squiggles, representing a mixture of different DNA fragments, each a different color. An arrow pointed to a surface where the fragments were dispersed as an array, so each could be sequenced as separate molecules. Balasubramanian calculated that this process would allow for a gaudy 1 million bases per run. His investors were skeptical, but said they'd be impressed if he could get to even 10 percent of that figure—which would still be a thousand times better than Sanger sequencing.[4]

In 1999, Abingworth gave Solexa a proper launch by putting up $3 million. Employee number three was Harold Swerdlow, a lanky, slow-talking sequencing expert from New York. The founding CEO was Nick McCooke, an engineer who had returned to the UK after a stint at a Seattle biotech. Two essential recruits who joined Solexa independently from Glaxo, medicinal chemist John Milton and bioinformatician Clive Brown, played crucial parts in making this process a reality. Instilling a pharmaceutical rigor that didn't previously exist at Solexa, Milton reinvented the chemical process piece by piece. "There's not a single atom left in the chemistry at Solexa that was there originally in the academic founders' labs," he told me. "They built a Model T Ford, but we had to build a Ferrari. It took four or five miracles for the thing to work end to end. Any one of them could've been a roadblock." McCooke allowed Milton to recruit dozens of top-notch chemists to improve the sequencing chemistry. "If the sequencing chemistry is no good, it doesn't matter how good the instrument is. You might as well be making cappuccino," Milton said.[5]

It is fair to say that Solexa's initial interest in personal genomics happened largely by accident. "We didn't go down to the pub and think about personal genomes as a goal," said Balasubramanian. "It was a scientific idea that seemed like a bit of fun. We almost feared finding a use for it." But McCooke had no such qualms, deliberately accentuating the medical implications. "Thanks to Nick, it was all about 'the $1,000 genome,'" said Brown. "We used that term in a lot of early spiels," agreed Swerdlow.[6] "Nick was always banging on me to get the cost down to $1,000." McCooke would love to claim credit for coining the phrase, but can't be certain. He did use it with investors in 2001, and also emphasized to the media Solexa's ultimate interest in personal genomics. In September 2002, he told the BBC that patients could soon expect to receive a complete map of their genetic code from their doctor which would be "kept confidentially with the rest of your medical records."[7]

In 2002, Solexa's Tony Smith gave a talk at Venter's symposium in Boston, but like everyone else, was overshadowed by Eugene Chan. "Every board meeting, the investors wanted to know—how are we doing relative to U.S. Genomics?" recalled Balasubramanian. Being British made a difference. "You have to struggle a bit harder to convince people you're worth backing, just because you're not in the U.S." he said.

In spring 2003, another competitor emerged 6,000 miles away when Steve Quake and colleagues at Caltech published a paper on a method for single-molecule DNA sequencing that was similar in some ways to what Solexa had been developing, only the Brits had chosen not to publish. Balasubramanian said he could have submitted the same story five years earlier: "Steve Quake was the first to publish [and] will say he invented single-molecule sequencing, but we had the first patent on it."[8]

By this time, the prototype Solexa sequencing system went like this: Spread a layer of short DNA molecules onto a slide, primed and ready for sequencing. Add the nucleotide mixture (the As, Cs, Ts, and Gs), each nucleotide tagged with a fluorescent probe and modified such that only one base could be added at a time. Once incorporated, take a picture of the fluorescent spots, unblock the bases, wash, rinse, and repeat. In practice, however, detecting fluorescent tags on single molecules of DNA was extremely hard because sometimes the fluorophores didn't light up. Brown remembered Swerdlow listing all the cumulative inefficiencies of single-molecule sequencing on a whiteboard. There was also a tendency to depict the DNA strands in slick marketing presentations as rigid, linear molecules, like a stack of pencils on a sheet of glass. "No, no, no!" laughed Milton. "DNA likes lying down. It doesn't stand up there like a telegraph pole. If you went down to the surface of the chip, it'd be like a mountain range in the scale of things. Some little bush of DNA hanging off and the enzyme's got to find it."

Klenerman had the interesting idea of putting a loudspeaker under the chip that would blast high-frequency sound waves to make the DNA stand on end. When that didn't work, McCooke faced two alternatives: build an enormous sequencer (the route that Quake's company Helicos took) or amplify each DNA strand and sequence hundreds of copies of each molecule, even if that meant abandoning Balasubramanian's original concept to sequence single molecules.

The safety-in-numbers approach looked even more appealing after McCooke heard about a Swiss company, Manteia, that had its own sequencing prototype and a neat method for amplifying the DNA strands into clusters of about 1,000 identical molecules. But it couldn't get the system working. After Manteia declared bankruptcy in late 2003, McCooke snapped up the cluster chemistry, which the Solexa team adapted and never looked back. It was a pivotal acquisition, for the bargain price of about $3 million, and a turning point in the Solexa story. "Buying Manteia instantly advanced us two years overnight. A fabulous acquisition," said Milton.[9]

Remarkably, Milton, Swerdlow, and Brown got the system working within a matter of weeks. Solexa managed to buy some of Manteia's instruments in an Internet auction on the cheap. "Our first sequence-by-synthesis data came off the Manteia machines," said Brown. "That's when we started sequencing properly."[10] Within three months, Solexa was cleanly reading 12-base DNA sequences. Six months later, it had doubled the read length to 25mers.

Three months after the Manteia acquisition, John West, a former Applied Biosystems (ABI) executive with deep knowledge of the sequencing market, assumed the CEO role from the popular McCooke. "The giant was asleep," said West of his former employer.[11] Solexa was running short of money, but Amadeus's Hermann Hauser led a new round of funding, contingent on West coming on board. West was worried about Solexa's long-term chances if it remained a purely British company. "We had great scientists, but no one for marketing," he said. Meanwhile, a Bay Area biotech called Lynx, the first commercial massively parallel sequencing business, was booking millions of dollars in revenue.[12] But Lynx's market cap and stock price were sinking, which hardly surprised Milton. "You can't run a sequencing business in California with all those West Coast salaries," he said.

During his first week on the job, West hatched a reverse-merger deal with Lynx CEO Kevin Corcoran. It looked fine on paper, but in reality it was a near catastrophe—what Brown termed "a massive merger bun fight" broke out over which company's technology would survive. Some Solexa board members initially favored the Lynx instrument, because Lynx supposedly had the better chemistry and engineering, a viable service business, a NASDAQ stock market listing, and an American base. Brown vehemently disagreed. "If you got through what actually was correct and true, it was that they had a stock market listing. The chemistry was dreadful, the instruments were useless, and the service business was dying or dead. They were three months away from bankruptcy." West took Brown's side, and told the board he'd rather make the Solexa chemistry work. "Apart from fund-raising, that's what I'd give him his top gold star for," said Brown.

Solexa's transition from a private British company to a publicly traded American firm was so stressful—the politics, the egos, the transatlantic culture clash—that Brown and others nearly walked. "Solexa was tightly run, no slack, hard-working. Lynx was three months away from everyone being sacked. It was ridiculous," said Brown. Every few weeks, Brown and Milton flew to San Francisco. Milton took charge of the chemistry, while Brown handled the software. On the engineering side, the Lynx prototype had a light source, a scanner, and fluidics—all the ingredients Solexa needed. But it was the British company's surface chemistry, hardware design, and software that won out. Corcoran and many of the Lynx engineers decided to return to ABI. "We had spent six months convincing them to build our machine, not theirs," said Swerdlow, who admitted the first machines "were pretty flaky." Still, by March 2005, Solexa was a publicly traded company on NASDAQ. "It was a stroke of genius economically," said Swerdlow. "Lynx was worth

$12 million, we were worth £20 million. Put it together, it very quickly became a $200 million company"—and one that was about to make some investors very happy indeed.

At last the Solexa team decided to try sequencing its first organism—the ΦX174 virus, the organism that Fred Sanger first deciphered twenty-five years before. Feeling confident, Brown persuaded a couple of colleagues to work over a weekend in January 2005, ordered in pizza, and started cranking the analysis software. Within a few hours and after some fancy math, Brown suspected that they had successfully sequenced their first genome and tapped out an e-mail entitled "We've done it." By Monday morning, he'd confirmed that, remarkably, the sequence was more than 99.9 percent accurate.[13] Within a couple of weeks, the ΦX174 feat had been repeated a dozen or more times. But neither Brown nor Milton was interested in pausing to write up and publish their result. "My interest is patents," said Milton. "We didn't publish as policy," said Brown. "It was a distraction." West had to be content with issuing a press release. The only dissenter was Swerdlow, who regretted not publishing the benchmark for others to cite later on. "It should have been Swerdlow et al.," he said. It was now "all hands to the pumps," and onto the next milestone: the human X chromosome.

Solexa wasn't unduly concerned when 454 released the first commercial next-gen sequencer in 2005. Milton suspected 454 would have problems, especially in sequencing stretches of so-called homopolymers: runs of the same base, such as five As or six Ts. "We used to call them '4, 5, or 6' as an internal joke, because they couldn't count repeats properly," Milton said.

Targeting a 2006 launch, West priced Solexa's 1G Genetic Analyzer (the 1G stood for the target output of 1 gigabase, or 1 billion bases, per run) at about $400,000. Over breakfast West sketched on the back of a napkin how the 1G machine would be able to sequence a personal genome for about $100,000 in three months.[14] But producing a complete genome sequence when each piece of the jigsaw puzzle was just 25–30 bases was a daunting computational challenge, even with the guidance of the reference genome.

For technical reasons, West's pledge of the $100,000 genome did not materialize in 2006, or in 2007 for that matter. "John said the end of the year, just not *which* year!" joked Swerdlow.

As commercial priorities shifted, Solexa grew and more and more sales and marketing personnel arrived. "It's fucking annoying, frankly," said Brown. "They didn't take any of the early risks. They come in with the fancy titles and the fancy packages. That was the trigger point for me deciding to leave." His software was working, he'd simulated a human genome assembly in the computer for the first time, and perhaps not insignificantly, his shares had vested. In mid-2006, Solexa did ten runs of the human genome, and the company bought expensive disk storage to handle the deluge of data—10 to 15 gigabytes per week, or five times as much as the Sanger Institute had at the time. "For about one year, we were the world's biggest genome center," Brown recalled.

By summer 2006, Solexa had placed its first instruments with genome centers including the Broad Institute and The Genome Center at Washington University, even though the instruments still had accuracy problems. West felt there was some denial going on, and following Milton's departure, set about fixing those problems. But Solexa also faced a dwindling cash reserve—less than $50 million—despite its sales, which would last only twelve months. At this point, West reached out to Illumina, a fast-growing genomics company in San Diego that specialized in genotyping and gene expression platforms. In November 2006, Illumina bought Solexa for $600 million. Illumina provided Solexa with an international sales force and support team, and Solexa turned Illumina into a truly comprehensive genome analysis company, the only company to offer instruments for analyzing gene variations, gene expression, and now DNA sequencing. Solexa's acquisition marked a defining chapter in what the Cambridge University Press office called "one of the greatest commercialisation success stories to emerge from the University of Cambridge." Illumina immediately sank $50 million into Solexa to further the commercial development of the Genome Analyzer (GA).

By February 2007, Illumina had placed forty machines in the field, with the Sanger Institute buying twenty-six. Although the read length was still a fraction of 454's, the throughput and cost per gigabase were all in Illumina's favor. As for its first human genome sequence, Illumina had produced threefold coverage, but that was not enough to start analyzing the data. Illumina CEO Jay Flatley told an analysts' meeting that he anticipated reaching a $100,000 genome by year's end. "A grad student will be able to sequence the genome of an organism as a PhD thesis," he said. Solexa's technology was driving the "democratization of sequencing," allowing small groups around the world to become their own genome centers. The traditional Sanger machines could produce 1 million to 2 million bases per day. By contrast, the GA could deliver 1 gigabase (Gb) over a three-day run.

Ironically, in the quest to complete the first human genome using Solexa sequencing, Illumina was scooped by one of its best clients. In October 2007, the Beijing Genomics Institute (BGI) in Shenzhen, China, announced that Jun Wang and his team had sequenced the genome of a Chinese individual[15]—just the third report of a personal genome sequence, following Watson and Venter. Chief scientist David Bentley finally completed Illumina's first genome, of an anonymous African individual, over six weeks around Christmas 2007. Performing twenty-seven runs, Bentley's team produced 77 gigabases of sequence, covering more than 90 percent of the genome.[16] The project uncovered more than 3.7 million SNPs, including 1 million novel variants, as well as all manner of structural variants and chromosomal rearrangements. The cost, according to company executives, was as West had predicted: $100,000[17]—a fraction of the cost of sequencing Watson or Venter.

Illumina's sequencing business exploded from zero to $100 million in one year. Its 2007 revenues doubled to $360 million as more than 200 GA instruments, with throughput steadily increasing, were installed by the turn of 2008. By year's end, that number had doubled again. The Broad and Sanger Institutes expanded their fleets, as did BGI, but it was the smaller labs driving most of the demand. An Illumina ad quoted one customer saying: "I can turn my mailroom into a genome center." In spring 2008, Illumina unveiled its second-generation instrument, the GA II (the original machine would henceforth be known as the "classic"). Improvements in hardware, software, and biochemistry upped the GA II specs to read lengths of 50 bases and a throughput of 750 million bases a day, 80 percent of them error free.[18] "We had a corporate goal in 2008 to make human genome sequencing routine," said Flatley.[19]

The Sanger's output in 2008 using Illumina machines was astonishing. In July, the institute celebrated its 1 trillionth base sequenced—1 terabase—the equivalent of 300 human genomes.[20] The Sanger press office helpfully pointed out that if the DNA sequence were printed in 12-point Courier type, it would span the earth sixty-three times. By the autumn, that data mountain had doubled. From September 2008 on, under the expert eye of Swerdlow (who had left Illumina), the Sanger was pumping out as much DNA sequence per week as in all of Genbank—the global gene sequence database. "Approaching 90 percent of all the DNA ever sequenced has been done with the chemistry I built and Clive's bioinformatics," said Milton proudly. (Both he and Brown joined another British sequencing start-up, Oxford Nanopore.) Brown agreed: "It's been phenomenally successful. I think we expected that. I think its John's chemistry. It's good, simplish chemistry on a pretty crappy instrument." It was also vital that Solexa reached the market ahead of ABI. Said Brown, "That's one of the few reasons Illumina dominates the ABI system—they got there first."

Solexa "could so easily have been a complete disaster," said Milton. "The Brits are the great inventors, and the Yanks are the guys who are great at putting it in a box and selling it." Rothberg may have declared victory in 2005, but three years later, Brown and Milton said categorically, "We won!" That view was echoed by many other sequencing experts and competitors.[21] "We're all very proud of it looking back, but God, it was a struggle," said Milton. "We didn't build the Solexa technology for people to go sequencing multimillionaires and giving them sequence that nobody understands, that they want printed out so that they can wallpaper their mansion. We built Solexa to sequence whole human genomes to better understand disease and cures and link that to the pharma industry."

Illumina sought to drive home its advantage when it unveiled its new top-of-the-line sequencer, the HiSeq2000, in early 2010. Some clever engineering tweaks had pushed the throughput in a weeklong run to 200 gigabases and the cost of a single human genome down to $10,000. Researchers could now sequence a patient's tumor genome and a control side by side in a week for just $20,000. The biggest endorsement of the technology came from BGI, which immediately placed an order for a staggering 128 machines, to be installed in a former printing press in Hong Kong. Flatley predicted that by the end of 2010, BGI would be producing as much sequence as the entire United States.[22] The Broad Institute followed suit by ordering 51 for its own fleet.

Balasubramanian was understandably elated with the performance and popularity of his sequencer, which had exceeded the performance specs he had scrawled out a decade earlier. He felt the key to the success hinged on putting quality first: "There were times when we could have rushed ahead, could have made claims ahead of performance. But we were very English, actually, and quite understated in how we put this out." There were some glitches at launch, but the machine had improved radically since then. "For me, what's been striking is the Moore's law improvements that have resulted from not changing the quality, but optimizing it and engineering it," he said. "The rate of change in performance is astounding." That said, "If someone can come out with something that's a thousand times better than what this can do, wonderful, because it'll force the revolution even faster. That's the scientist in me." The goal of a $1,000 genome in one or two days was tantalizingly close, possibly arriving as early as 2010, he reckoned. "If it's not Solexa sequencing, it's going to be something else. It's going to happen."

• • •

One of the biggest incentives driving the race for the $1,000 genome was a $10 million purse, the Archon X Prize for Genomics, to be awarded to the first group that successfully sequenced the genomes of 100 individuals in ten days for less than $10,000 apiece. The idea for rewarding a breakthrough in next-generation sequencing began with Craig Venter, who announced his own $500,000 cash prize back in 2003. He later agreed to roll that into the Archon X Prize, after Google's Larry Page invited him to join the board of the X Prize Foundation (which had awarded its inaugural prize—the Ansari X Prize—to Burt Rutan for designing the first privately funded space plane in 2004). X Prize officials pointed to the example of Charles Lindbergh, who flew across the Atlantic to claim the $25,000 Orteig Prize in 1925, as evidence of the value of such prizes in spurring technology development.

The benefactor of the genomics prize was the unlikely figure of Canadian diamond hunter Stewart "Stu" Blusson, a private individual who would rather be camping in a tent than hanging out with high society or talking to the media.[23] To my surprise, when I called the Vancouver office of his company, Archon Minerals, it was Blusson who answered the phone. He explained he ran a one-man company and couldn't be bothered to build a Web site or hire a receptionist.[24]

Blusson spent twenty-five years conducting geological surveys across Canada, surviving helicopter accidents and bear attacks, before teaming up with fellow geologist Charles Fipke in 1979 to search for diamonds, competing against the industrial muscle of De Beers. In 1991, Fipke discovered a diamond cache just 150 miles south of the Arctic Circle. In code, Fipke would call his team, "How's the fishing?" The crew chief replied, "Chuck, we just caught the biggest fucking fish you ever saw!"[25] That haul became the Ekati diamond mine, which yields five million carats of gems a year. A portion of Blusson's newfound wealth is now the Archon X Prize, named after the Archean Craton, the ancient geological plate that lies beneath Canada. "Here's a real opportunity to change medicine for everybody. It's incredible where this could lead," he said.

That said, the rules of the competition are daunting. Crucially, the finished sequence must cover at least 98 percent of all twenty-three pairs of chromosomes (that is, the diploid genome) at a stringent accuracy of 99.999 percent or just 1 error per 100,000 bases. To date, no human genome sequenced has matched these criteria at any price. "I think [that] will be impossible for decades and decades," said Jonathan Rothberg. "I'd love to go for it, but it's impossible with that linkage requirement."[26]

To sweeten the pot, the X Prize Foundation dangled a further $1 million for the winner to sequence the genomes of 100 specially chosen volunteers. Among the early volunteers were Microsoft cofounder Paul Allen, CNN's Larry King, and astrophysicist Stephen Hawking. (One blogger, noting the male-heavy roster of DNA donors, wrote sarcastically that perhaps they should call it "the Y Prize.") Hawking said he hoped the prize would "help drive breakthroughs in diseases like ALS [Lou Gehrig's disease] at the same time that future X Prizes for space travel help humanity to become a galactic species."[27]

Perhaps because of the stringent criteria, the initial reaction to the bounty was a little muted. By 2010, only eight groups had officially registered for the competition.[28] "We didn't start the company to win the X Prize," said Helicos Biosciences founder and chairman Stanley Lapidus,[29] who chose not to enter, preferring to focus on capturing a slice of the $1 billion to $2 billion genome-sequencing market. Hugh Martin, CEO of Pacific Biosciences, guaranteed his firm would win the X Prize "when we're ready"—even though it had yet to sign up.[30] Meanwhile, back in the UK, no one could top the hubris of Brown and Milton's new boss, Oxford Nanopore CEO Gordon Sanghera. "Oh, we'll win the X Prize," he told me, "then refuse the money!"[31]

CHAPTER 6

Service Call

While Illumina had captured the lion's share of the next-generation sequencing market by the end of 2009, a rapidly growing list of competitors were snapping at its heels. Some, like Helicos Biosciences, were pioneering new methods of sequencing, pushing the envelope of detection by focusing on single molecules of DNA. Others, like Complete Genomics, had a bold new business model that forsook selling machines in favor of delivering information—the finished sequence. Meanwhile, Applied Biosystems (ABI), the company that had enjoyed a virtual monopoly in the sequencing business throughout the 1990s, even supplying both sides in the race for the Human Genome Project (HGP), reacted aggressively to recapture its previous dominance.

ABI's counterinsurgency plan necessitated acquiring a next-generation sequencing company. After studying more than 40 prospects, the company decided that Agencourt Personal Genomics (APG) was the best choice. APG had made good progress with a new process of sequencing that used a technique for gluing strands of DNA together rather than synthesizing new strands, using enzymes called ligases, that bond the fragments together. APG became the crux of what ABI dubbed the SOLiD system—for Sequencing by Oligonucleotide Ligation and Detection.

In spring 2007, the head of the APG effort, Kevin McKernan, took me on a tour around the ABI production facility in Beverly, Massachusetts.[1] It wasn't so much a facility as a windowless room containing about ten SOLiD prototypes, their innards exposed for technicians to tweak before the plastic-coated shells arrived. He seemed unconcerned that Illumina and 454 had been selling instruments for eighteen months. "This is not a race that's going to be over in six months," he said. "It's a race that's going to go on for years—it's a marathon." As scientists fussed over some of the machines, I noticed that a wooden nameplate was attached to each instrument. One said Amelia, another Barbara, and a third Rosalind. McKernan explained that each instrument was named after a famous female scientist: Amelia Earhart; Nobel laureate Barbara McClintock; Clara Barton, founder of the Red Cross; Florence Nightingale; Rosalind Franklin; Junko "Joan" Tabei, the first woman to climb Mount Everest; and Liz Blackwell, the first American female doctor.

It was ironic that ABI should pin its hopes on McKernan, a man who had spent the ulcer-inducing years 1999–2000 working night and day with Eric Lander at the Whitehead Institute genome center, spearheading the public consortium effort against ABI's sister company, Celera.[2] The approach that McKernan devised at APG was an interesting mix of novel and what he called "recycled" technology, some licensed from Harvard Medical School's George Church. The novel ideas came from Alan Blanchard, a smart Caltech-trained scientist who had formerly worked at Manteia and had interviewed unsuccessfully at 454 before joining APG in late 2004. McKernan took Blanchard to lunch on his first day, and the pair began sketching ideas that became known as "two-base encoding." Put simply, the method reads each base twice, providing greater accuracy than competing systems.[3] Two years later, ABI decided to place a $120 million bet on APG's next-generation sequencing technology.

ABI officially released its SOLiD instrument for sale in late 2007. Michael Rhodes, a British product manager at the company, told me shortly thereafter that the SOLiD system would perform like "a Formula One car."[4] I told him I thought a NASCAR analogy might appeal more to an American audience. "Nah, Formula One's faster," he replied. Two months after Illumina's announcement that it had fully sequenced a human genome for about $100,000, ABI trumped Illumina with a press release claiming it had completed its first human-genome sequence at a cost of a mere $60,000 (not including staff, overheads, depreciation, or nameplates). Science by press release doesn't have the credibility of peer-reviewed articles of course, and the announcement merited an asterisk. But McKernan's group said it had produced thirty-six gigabases of sequence data (or twelvefold coverage) in seven SOLiD machine runs and had reached 99.95% accuracy. McKernan's group transferred the sequence data to the National Center for Biotechnology Information (NCBI), where the files were made publicly available.[5] That summer, using four machines, running over two months, it generated more than 250 gigabases (Gb) of sequence—more than all the data stored at the time at Genbank. (ABI's first human genome using the SOLiD sequences was published in 2009.)[6]

ABI began aggressively pushing SOLiD instruments into the marketplace, and launched a new ad campaign that asked: "What would you do with the $10,000 genome?" It placed five more SOLiD machines with the Broad Institute, but then the company was dealt a setback. The Wellcome Trust Sanger Institute in the UK decided to return their machines and go with Illumina's instead. The Sanger said it had given the SOLiD machine a full evaluation, but in the end concluded that growing its fleet of Illumina instruments was more practical. McKernan, aggrieved by the Sanger Institute's decision, vented his frustration in an extraordinary three-page letter to the House of Lords, which was conducting an inquiry into personalized medicine.[7] McKernan wrote that he had once been a teammate working alongside "the Brits at the Sanger Institute to fend off the threat of a genome monopoly" by ABI and Celera. He alleged that the Sanger Institute's decision to commit to Illumina smacked of a conflict of interest and threatened the UK's "competitive positioning" in genomics. "Our simple presence in their lab will ensure Illumina is always providing a competitive price," he wrote. By the end of 2009, the Broad Institute had also made its bed with Illumina, upping its fleet of Illumina's machines to a whopping 89, compared to a handful of SOLiDs.[8] The Genome Center at Washington University, which published the first cancer genome, had also returned its SOLiDs and was banking on Illumina.[9]

In 2010, Life Technologies (ABI's new parent company) began clawing back some momentum and market share, led by a partnership with Dietrich Stephan to furnish the Ignite Institute of Individualized Health with no fewer than 100 SOLiD sequencers. Thanks to continued enhancements from McKernan's team, the new instruments were on target to deliver 300 Gb sequences per run by the end of 2010, which would immediately catapult Ignite to the top of the genome center charts. As importantly, the cost target was just $3,000 for a human genome, at an accuracy of 99.99 percent. And waiting in the wings was a promising third-generation system based on two other biotech acquisitions, imaging single molecules of DNA.

"There haven't been a lot of revolutions in medicine," says Stanley Lapidus.[10] "Medicine is mostly a story of no progress at all until about 1800. It is vaccination, anesthesia, infectious disease, microscopes, the X-ray tube, and the genomic age. That's it, that's everything. There's nothing else! A tremendous amount of incremental progress, but those are the revolutions." In founding Helicos Biosciences, Lapidus was determined to add to that list.

In 2003, Lapidus was moonlighting as a nature capitalist when he stumbled on a paper by Caltech professor Stephen Quake describing a primitive technology that could barely sequence five bases of DNA.[11] That paper wasn't published in a trendy journal, but Lapidus was convinced it could provide the foundation for a device he later dubbed the HeliScope. He believed it could do for cancer and heart disease what the microscope had done for typhoid and tuberculosis. The machine Lapidus would build would take images of billions of single molecules of DNA, allowing for greater speed and accuracy than other next-generation techniques.

In pitching his company, Lapidus would don his trademark bow tie and put on a professorial demeanor, laying out a bold vision. "Americans lived to about age forty through the early parts of the twentieth century," Lapidus explained at one presentation. "The greatest triumph in increasing life span was the invention of the microscope. This was the beginning of the life-sciences tool industry as we know it. Within a generation, it had tremendous impact on reducing and ultimately eradicating mortality from the most common forms of infectious disease. It resulted in almost a doubling of our life span."[12]

At this point, Lapidus showed a slide comparing the three major diseases of the nineteenth century—cholera, typhus, and tuberculosis—to the scourges of the twenty-first century—heart disease, type 2 diabetes, and cancer. "Today, we die of cancer, heart disease and its sequelae," he said. "These are diseases that are essentially genetic diseases—genes we pass from generation to generation, or diseases that begin with the genetic alteration of a single cell, as in cancer." Just as the light microscope solved the mysteries of infectious disease, Lapidus argued, his "DNA microscope" would conquer the diseases caused by the complex interplay of genetic and environmental factors.

He ended by giving a taste of the planned specifications and performance of the HeliScope. While the first instrument would sequence about 100 million bases per hour, Lapidus said, the ultimate capacity of the instrument would be 1 billion bases per hour. You didn't need to be an investment banker to do the math. "We're talking about genomes per day," he said. "The trajectory of improving our chemistry will take us to a $1,000 genome and below." Leading the technology development were two industry veterans: Bill Efcavitch, who had masterminded ABI's automated Sanger sequencers throughout the 1990s, and Tim Harris, who had worked at SEQ with Kevin Ulmer. On his first day at Helicos in 2004, Efcavitch wrote down his ideas for the commercial specifications, which he put in an office drawer. The top priority was to build a machine that could sequence a human genome tenfold in twenty-four hours.

Whereas Solexa had abandoned the idea of imaging single DNA molecules, Helicos believed it could develop the first "true single molecule sequencing" system. The idea was to break DNA into short strands about 200 bases long, then attach these molecules to a slide using a kind of molecular Velcro, producing a forest of about 100 million DNA molecules per square centimeter. After fluorescently tagged bases had been added, the molecules were illuminated by firing two watts of laser power—the equivalent of 1,000 laser pointers. A camera snapped a flurry of photographs with 50-millisecond exposures, each image capturing a tiny portion of the slide containing about 40,000 DNA strands, for a total of 90 million bases per hour. Capturing 100 nanometer pixels at fifteen frames a second necessitated the camera to settle down to literally zero vibration after each step. For that, Helicos housed the apparatus in 300 pounds of New Hampshire's finest granite. In all, the towering HeliScope weighed in just 20 pounds shy of a ton.

I first met Lapidus in the summer of 2007, just weeks after he'd taken Helicos public. He'd been through a grueling month of presentations that had netted the company about half of the $100 million he had hoped for. But he didn't appear concerned and exuded complete confidence that the HeliScope would transform the race for the $1,000 genome. If Helicos could compete with 454, Illumina, and ABI, it would indeed be a historic company, although it would have to go some to match Lapidus's earlier success—Cytyc, the developer of the thin-prep pap smear test—which "went from my basement to a $6 billion acquisition in twenty years." Lapidus explained that the goal was still to build a machine that could sequence a human genome tenfold in twenty-four hours.

If Lapidus was concerned about Helicos's late arrival to market, he wasn't letting on. "Do you remember what the first computer company was?" he asked me.[13] "I've no idea," he continued, without pausing. "It's like computers in 1954, where the race was on with the vacuum tubes and the transistor." According to Lapidus, Illumina and ABI, which both relied on the DNA amplification process, were the sequencing equivalent of the vacuum tube. "It is fine we're not the first in the market. We believe we have the best platform and the best trajectory," Lapidus asserted. "As we move to the $1,000 genome, that means it's a mass market. The Helicos advantages will mean that Solexa and Illumina will be footnotes in history." Lapidus paused for a second to reconsider. "Let me say that in a nicer way. It's inconceivable that ten years from now anything other than single-molecule technology will be used for nucleic acid measurements. Maybe it isn't Helicos, but I believe it will be."

When I mentioned that 454 had garnered enormous publicity by sequencing ancient DNA, Lapidus almost spat out his water. "I didn't start this company to do the Neanderthal!" he sputtered. "What unites us as a team is this idea that the simplicity, the speed, and the cost of what we do really unlocks the era of genomic medicine. It's great to do agricultural applications. It's great to understand the evolution of mankind. It's great to understand one's own family tree. These are all wonderful things. But it's really about medicine. It's about reducing morbidity and reducing mortality, living longer and better lives."

Whether Helicos will ever accomplish Lapidus's goals remains to be seen. The single-molecule sequencing method the HeliScope is based on has proved extremely difficult to build into a machine at a competitive price.

Lapidus had hoped to ship his first HeliScope before the end of 2007, which didn't happen. A *Wall Street Journal* story speculated that the price tag could reach $2 million, four times the cost of Illumina's GA instrument.[14] Internally, Helicos business development director John Boyce was lobbying for a price closer to $500,000, based on analysis suggesting there was a ceiling for the amount of government reimbursement purchasers would be entitled to. Lapidus ultimately priced the instrument high, at $1.35 million, and potential buyers complained of sticker shock.

In March 2008, Helicos finally placed its first HeliScope, with a biotech services company in North Carolina called Expression Analysis. Efcavitch stated publicly: "I don't want to be overly arrogant, but we see a clear technological pathway to a $1,000 genome. I think it's a two-year event."[15] But Helicos would have to overcome several vexing technical challenges, not merely in generating read lengths greater than thirty to fifty bases, but in solving the persistent problem of "dark bases"—the incorporation of bases that failed to fluoresce under the laser spotlight. The company's stock price, which had doubled after a successful public offering, sank below the launch price of eight dollars. Help arrived when two of the firm's cofounders, who were both running research efforts, ordered machines. Quake announced that he was collaborating with some fellow Stanford faculty to acquire an instrument,[16] and Eric Lander of the Broad Institute also ordered one. But Helicos was burning $2.5 million a month, and in late 2008, was forced to lay off 30 percent of its workforce.

When Efcavitch gave scientists an update on Helicos's progress in early 2009, he opened by saying, "The rumors of our demise are greatly exaggerated."[17] When an Irish geneticist stood up and asked why the HeliScope was so expensive, Efcavitch bristled. "We just announced a price reduction!" he said. "The Large Hadron Collider is an expensive instrument. We're much, much below that!" Just when the company's prospects seemed to be dimming, however, a silver lining appeared. In a stunning op-ed column in the *New York Times*,[18] Quake revealed that he had sequenced his own genome on the HeliScope at Stanford in March 2009, partly in hope of learning why his daughter suffered peanut allergies. Publication of Quake's genome attracted great attention for the number of coauthors on the paper (just three) and the cost of materials (under $50,000).[19] Whether it would be enough to keep Helicos in the $1,000 genome game or relegated to more of a niche role remained to be seen.

• • •

Another company is taking a radically different approach to the race to the $1,000 genome, banking not on selling fancy machines but rather on building a sequencing factory to sell sequence—the Netflix of next-generation sequencing.

On a particularly frigid night just before Christmas 2008, in Mountain View, California, a group of homeless people lit a bonfire that quickly spread out of control, engulfing a nearby electrical power pole. The fire department asked the electrical company to shut off the power, which might have wreaked havoc at a nearby business conducting a crucial experiment. At Complete Genomics, in the tropical confines of the "Hawaii Room," two rows of robotic instruments were whirring away, sequencing a human genome using a novel method involving billions of DNA "nanoballs."

Thanks to a backup generator, Complete's machines kept running. The control system monitoring the machines got a little confused, however, and the facilities manager had to scramble up to the roof in the rain to make repair to it. "This was an important genome for us and we didn't want to restart it," said information technology chief Bruce Martin.[20] The incident was an inadvertent demonstration of the resiliency of the new sequencing technology.

Just ten weeks earlier, in October, Complete had emerged from stealth mode to shake up the genome sequencing field. CEO Clifford Reid had declared that his mission was to be "the global leader in complete human genome sequencing,"[21] to transform the economics of genome sequencing by building a commercial sequencing factory to do "diagnostic quality human genome sequencing at a medically affordable price." He wanted to transition genome sequencing from a purely research endeavor into a mainstream pharmaceutical and medical enterprise. Complete would offer genome sequences at rock-bottom prices—as little as $5,000 initially—while sparing clients the hassle of running and maintaining their own instrument fleets and data centers.

What was extraordinary about Complete wasn't so much the underlying technology, which bore similarities to methods championed by both ABI and George Church, but Reid's business model.

Rather than sell $500,000 instruments, Complete would be a wholesaler of human genome data to anyone: Big Pharma, research institutes, and eventually perhaps direct-to-consumer firms such as 23andMe. Reid's audacious plan called for the construction of a $75 million sequencing factory in Silicon Valley running 96 instruments by 2009, which would churn out 1,000 human genomes that year, and by 2010 would sequence an astonishing 20,000 genomes with 196 machines. Within five years, Reid predicted, Complete would build ten genome centers across the globe at $50 million a pop. The plan was to partner with foreign governments, such as Japan, India, and China, which forbade the export of DNA samples. All told, Complete would sequence a staggering 1 million genomes within five years—the equivalent of 1,000 people with each of 1,000 diseases. "By the time we've done that, we will understand the genetic basis of all the important human diseases," Reid bragged.[22]

A highlight of Complete's emergence into the public eye was Reid's revelation that his firm had already assembled a rough human genome for the absurdly low cost (in materials) of just $4,000, and had done so by running just four instruments for a week, or as Reid put it, a "twenty-eight instrument-day experiment." Complete's technology was ten times cheaper and ten times faster than its rivals, said Reid, but that wasn't the half of it. By 2009, he predicted his company would be selling complete human genome sequences for just $5,000, including material, instruments, and labor costs, as well as overhead.

Reid saw two essential advantages in his services model. A huge untapped market was Big Pharma, which would jump at the chance to sequence the DNA of patients enrolled in clinical trials. That would allow the companies to distinguish drug responders from nonresponders in the trials or to identify genetic predispositions that patients might have to adverse effects of drugs. The pharmaceutical companies were now outsourcing more and more of their drug development operations, so why would they want the headache of building and maintaining their own sequencing operations if they could buy genome data wholesale? A second rationale concerned the ridiculous volume of data produced by sequencing. Complete was building a Google-style data center, matched only by, well, Google, and a handful of other high-tech organizations. Complete's data center was being built in a warehouse in Santa Clara, California. The company would relieve clients not only of the need to run their own sequencers, but also of the need to store their data. I suspected that a third, unstated, rationale for the service model was that the molecular wizardry behind Complete's sequencing method was awfully complicated, not the kind of simple recipe that could be routinely handled by labs around the world.

Reid anticipated luring pharmaceutical companies studying cancer and mental illness as early adopters and was also eying the consumer-genomics community. "Knome and 23andMe and Navigenics and all those guys will essentially buy genome services from us,"[23] he predicted. Naturally, Reid also courted genome research centers as early partners, as validation of Complete's model, even though the service might seem to run counter to the center's self-interest. The Broad Institute's Eric Lander signed an agreement to sequence the genomes of five kinds of cancer at $20,000 per genome. But some researchers were critical. One furious cancer researcher told me: "How am I going to get the NIH to fund my next sequencing grant with Complete Genomics claiming they can sequence a $5,000 genome without presenting a shred of evidence?" Meanwhile, one of Complete's competitors was whispering that the only reason Reid made his public announcement was that he was running out of cash.

A couple of weeks after Complete emerged from stealth mode, Jim Hudson, a biotech entrepreneur turned philanthropist and the founder of the Hudson Alpha Institute in Huntsville, Alabama, ran into two of Complete's executives attending a conference at Cold Spring Harbor Laboratory. He had heard about Complete's "$5,000 genome" through a Google Alert e-mail. "I love that thing," he said, in his leisurely Southern drawl.[24] As the founder also of the hugely successful biotech firm Research Genetics, Hudson knew a thing or two about running a profitable service business for the genomics community. If Complete really could sequence a genome for $5,000, Hudson thought there wasn't much point in his scientists doing the sequencing themselves. As luck would have it, Hudson happened to have on his person $6,000 in new bills. Was that typical walking-around money? "Well, sometimes," he laughed. "I didn't feel good leaving it back in the room, so I was carrying it in my pocket." Hudson introduced himself to the Complete staffers and asked them if they were serious about the $5,000 genome. "Are you really gonna do that?" Sure, they said. So Hudson pulled out his wad of crisp $100 notes and offered them $5,000 in cash. "It was kind of amusing," he said. "They backed up and didn't want to take it. But I did get their attention."

Complete's October surprise did not come as news to Kevin Ulmer. Since launching SEQ, the first single-molecule sequencing company, two decades earlier, Ulmer had studied and deconstructed every conceivable sequencing technology and had consulted for Helicos for nearly two years. Ironically, he had come full circle and had founded his latest start-up, Genome Corp., a DNA sequencing factory based on, of all things, Sanger sequencing. "You keep pushing the boulder up the hill, and Sisyphus eventually gets there," he said.

Ulmer had been born again. He knew the intimate strengths and weaknesses of every other technology and believed there was a huge upside to reinventing Sanger sequencing on a grander scale. "Sequencing is an information-services business," he said, sipping an Arnold Palmer overlooking Cohasset Harbor.[25] "What scientists want—whether an anthropologist, an ecologist, or a clinician—is sequence information, not another expensive box." With some modest seed funding and a small laboratory, Ulmer was laying out plans to switch on his sequencing production line in 2010. That was until he met Complete's founder Rade Drmanac three months before Complete came out and learned they were both "singing from the same hymnal." To make matters worse, both wanted to build a sequencing service focused on humans. "It's the only genome that makes sense to sequence millions and millions of times," said Ulmer.

Ulmer signed a term sheet with a Boston venture capitalist, but Complete had a three-year, $35 million head start on Ulmer, and the global financial meltdown struck the final blow. Because he couldn't beat Complete, Ulmer joined them as a consultant. In the face of another promising idea come undone, Ulmer was philosophical: "If someone would just pay me *not* to start new ventures, the global economy wouldn't collapse!"

It would have been easy to dismiss Reid's brazen claims of a $5,000 genome except for the reputation of Rade Drmanac, the scientific brains behind Complete. Drmanac has a large, square head and a thick mop of graying curly hair. He left his home of Belgrade, in the former Yugoslavia, in the late 1980s, but still speaks in a thick Serbian accent, with heavily rolled "Rs," accentuated consonants, and the occasional missing preposition. George Church, a Complete board member, who has known Drmanac for two decades, paid tribute to Drmanac's stubborn persistence by quoting George Bernard Shaw: "Reasonable people adapt themselves to the world. Unreasonable people attempt to adapt the world to themselves. All progress, therefore, depends on unreasonable people." Drmanac, said Church, "is unreasonable."[26]

As a graduate student back in Belgrade in 1986, Drmanac found isolating human genes a tedious job. Much like Craig Venter, he wanted to sequence an entire genome instead of painstakingly identifying one gene at a time. "That was my dream and my blood,"[27] he told me. "Maybe I'm lazy!" Drmanac sketched out his idea for a new method called sequencing-by-hybridization (SBH) shortly before moving to London. Any piece of DNA could be broken down into its constituent pieces. By making all possible combinations of primers and observing which ones hybridized to the target DNA, it should be possible to piece together the original sequence. In 1989, he outlined his idea in a scientific paper, but it nearly didn't see the light of day. Drmanac said, "I really owe my career to Victor McKusick," the legendary Johns Hopkins medical geneticist who had launched a new journal called *Genomics*, to which Drmanac submitted his manuscript. The reviewers of the paper were split, but McKusick elected to publish anyway. "If he was there, I would kiss him!" said Drmanac. "Without that paper, I would not be able to graduate."

After a stint at the Imperial Cancer Research Fund in London, Drmanac and a small group of fellow researchers took up positions at the Argonne National Laboratory in Illinois. In 1993, *Science* magazine published his most important paper to date, which offered proof that SBH worked.[28] With the backing of $500,000 from his friends, Drmanac then started his first biotech company, called Hyseq, building huge databases of the activity of thousands of genes in scores of normal tissues.

By 2003, Hyseq, which is now named Nuvelo,[29] was finally on a path to profitability, having reinvented itself as a drug development company. Drmanac wasn't interested in that line of business, and in 2004, Nuvelo predictably sold off Drmanac's genomics research part of the business. Drmanac was determined to be the buyer. "I have no money," he told his employers. "I cannot pay anything, but I can *promise* to pay you." Drmanac persuaded Nuvelo to loan him the business in exchange for $1 million to be paid within four years or Drmanac would have to return his patents. "It was a great deal—I cannot lose! If I'm successful, I pay them. If everything fails, I just return failed patents."[30]

Drmanac's meeting with Clifford Reid came quite by chance. Drmanac was looking for a chief financial officer; Reid had been walking the halls back at his alma mater, MIT, increasingly drawn to the prospect of building a biotech business. Drmanac and Reid hit it off immediately when they met, and within half an hour, Drmanac realized Reid understood his methods and was asking all the right questions. "I checked his CV: undergrad MIT, MBA Harvard, PhD Stanford—and a nice guy. Maybe there is a better guy than him as

CEO, but my chance to find him is absolutely zilch."[31] Reid did not know much about genome sequencing, but Drmanac's idea was undeniably exciting.

Initially, Drmanac and Reid left the question of how their company would make money open. Reid wasn't terribly concerned. In fact, he didn't even write a business plan. Ten PowerPoint slides sufficed to describe the business to prospective funders. By 2006, Drmanac was perfecting a new sequencing recipe. His approach resembled the ligation methods developed by Church and licensed to ABI, but featured several innovations. Drmanac divided the human DNA to be sequenced into lengths of seventy bases. He then punctuated those stretches by inserting short DNA adapters to serve as zip codes or signposts, from which each stretch of DNA sequence could be read.[32]

Another critical innovation was the development of a gridded array, similar to 454's, but with much smaller wells. A standard slide could hold up to 1 billion wells, each just 300 nanometers in diameter and spaced 1 micron apart. They were etched into the slide using a process called photolithography. Drmanac created a solution containing some 10 billion DNA nanoballs per milliliter, and spread them across the slide, filling about 90 percent of the wells. Complete could in principle generate up to 70 billion bases of sequence per slide this way, enough to sequence an entire human genome in about a week.

Reid and Drmanac agreed that the business should resolutely sequence nothing but human genomes. Drmanac therefore didn't even bother sequencing the usual milestone organisms such as bacteria or fruit fly. Some board members worried that Drmanac was being unrealistic, but he knew his technology worked, and he set a deadline of just two years for achieving commercial viability. Complete's first effort to sequence a human genome took place in the summer of 2008, with mixed results but good enough that Reid could talk about the attempt when Complete emerged out of stealth mode that fall.

The run that was nearly interrupted by the fire just before

Christmas 2008 was a good deal more successful. "This is probably the best sequenced genome from the coverage point of view ever," he said.[33] However, with an error rate of 1 in 10,000 bases, the results would still not be good enough for personal genomics; he was shooting for 1 mistake in 10 million. Drmanac took me on a stroll through the Hawaii Room a few weeks after they had finished that run. Walking down the long, windowless lab bathed in fluorescent blue light felt like walking into a nightclub, albeit without the music. Blue light, Drmanac explained, falls on the far left of the visible light spectrum, so it doesn't interfere with the cameras snapping photos of the DNA nanoballs. The high humidity prevents the evaporation of the solution from the walls in the slides, which are open to the environment. As Drmanac noted, "it's better than working in the cold room."

On both sides of the room, sequencing machines were humming, sixteen in all, sequencing genomes for Lee Hood of the Institute for Systems Biology. I permitted myself a slight smile: Hood had praised Complete on its launch even though he admitted to me later he hadn't yet seen any of the company's results.[34] The cost of each machine would be about $200,000. Some visitors to the Hawaii Room were reportedly disappointed—the sequencing instruments aren't much to look at—but Reid's view was that it's not about the instrument but the biochemistry.[35]

The first machines sequenced twenty-four hours a day, seven days a week. The early output was twenty genomes per month, but for commercial viability that production would need to grow tenfold by 2010. Even that wouldn't satisfy Drmanac, though. Complete's first genome was sequenced on a single instrument over ten days, but Drmanac wanted to do ten genomes in one day. In other words, he wanted to improve Complete's results by a factor of one hundred within two years, relying on improvements in engineering, imaging, and chemistry.

In February 2009, Reid gave his first major talk about Complete's results at a conference on Marco Island, Florida. Sensing that there were some people rooting against the company, Reid dispensed with formalities and announced that Complete had already sequenced its first human genome. "Not an interesting individual," Reid said condescendingly, meaning the genome was not that of a celebrity scientist but of an anonymous male Caucasian HapMap donor. George Church had examined the data, Reid explained, and had said it looked solid. He had called the sequence "a major achievement" that "surpassed expectations."[36] Reid said he asked Church for his opinion "just to make sure we weren't smokin' dope."

According to Reid, the cost of materials for Complete's first genome was less than $4,000, and that figure was shrinking fast. But he hedged on his original pledge to offer $5,000 genomes by spring 2009. The unspoken reason was that Complete was having trouble raising another round of venture capital. The company couldn't afford to sell every genome at a rock-bottom price, and the revised pricing structure listed whole genomes at $20,000 each, dropping to $5,000 only for bulk orders. "Send us your samples," he told the Marco Island crowd, so long as they were human genomes. "We'll sequence them, we'll assemble them, we'll generate the variants list, and we'll send it back to you quickly."

Reid had leased an additional 32,000 square feet of lab and office space to handle a fleet of new commercial-grade instruments. By 2010, Complete's data center would be half the size of the largest computing center in the world. The task of building and managing that behemoth, capable of handling thousands of genomes a year, belonged to Bruce Martin, one of the original architects of the Java programming language. Martin recruited a tranche of Silicon Valley experts in high-performance computing and data mining. By 2010, Martin expected to be storing about 30 petabytes of data.

The day after Reid's talk, Drmanac and I talked about the possibilities in personal genomics while enjoying some Florida sunshine. Personal genome sequencing "is a universal genetic test,"[37] he said. Complete could become the wholesale provider of individual genomes for 23 et al., he insisted "but for medical purposes, not for fun," and not just for the wealthy. "You have to educate people upfront. The most likely answer when your genome is sequenced is, 'We didn't discover anything to report to you, so behave as you would behave.' Only one in ten people will have a report that says you should do something different. People should be happy that you are in that group. . . . You paid this, I got nothing? No, you got your genome, you're a lucky guy."

Drmanac says the odds ratios of say 1.2, which are commonly reported for the odds that specific genes are implicated in common diseases, don't mean much by themselves. "I may be two times less likely to contract a disease," he says. "1.2 is an average for that group of people. Nobody knows my chance. It's calculated for that SNP, not my genome." If the result of a personal genome scan is that "we cannot tell you anything useful, I'll be happy with that," he continued. I asked Drmanac if he was excited or concerned about Reid's pledge to sequence 1 million genomes by 2013. As it happened, he was neither. "There are 4 million babies [born annually] in the United States. So we need 4 million in a year, not 1 million in five years. So we're happy in one way, but this is nothing."

For many years, the cost of DNA sequencing fell by 50 percent every twelve months. But around 2005, the price began dropping tenfold every year. The price of genome sequencing was collapsing. Extrapolating from that trend, the cost of genome sequencing in 2015 would be about one cent. "This will not happen, I assure you!" said Reid.[38] "What you're seeing here is not a Moore's law phenomenon, not an incremental improvement on an existing technology, but rather the entry into the genome-sequencing world of a series of disruptive genomic technologies."

Reid predicted that by 2010 he would be able to read a complete human genome on a single slide in half a day. "It's an absolutely extraordinary cost point for human genome sequencing," he said. As the cost of materials approached zero, because of the improving miniaturization and parallelization, the cost of sequencing could be boiled down to the costs of imaging and computing. By centralizing the data processing, Complete would afford scientists the luxury of focusing on the science and medical research. "We're able to rely on the two formative distribution industries of the modern era: FedEx, for moving atoms, and the Internet, for moving bits. The model is: Send us the sample, we'll sequence it and send you back the data. Don't worry about the industrial engineering that went on in between there. That's a back office function."[39] He said, "There's one species worth sequencing millions of times, and it's in this room." And by "worth sequencing," he meant profitable.

In a similar vein to Stanley Lapidus, Reid described his sequencing setup at one conference as a "gene microscope."

> The light microscope was invented in 1590 by a father and son team—the Janssens—a couple of opticians in the Netherlands. The light microscope changed science. The scientific method is hypothesize, test. Well, measurement tools are the test part of that. Whenever you see an advancement in measurement tools, there's an advancement in science. The light microscope advanced botany (leaf structures). It advanced geology (crystal structures). . . . But there was one noticeable part of human endeavor it did not advance at all—and that was medicine. The light microscope of 1590 wasn't really good enough to look inside a cell and see what was going on. You could see cells—Robert Hooke coined the term *cell* in 1656—but very little happened in medicine for 300 years.
>
> Then in the 1870s, there was a revolution in light microscopy. Some guys invented oil immersion lenses, the Germans invented aniline dyes, and for the first time, in 1879, you could look inside a cell and see what was going on. The cellular cause of tuberculosis was discovered in 1882. TB at the time killed more people on the planet than any other disease. There was more advancement in medicine in the three years of the "good" light microscope than the 300 years of the "bad" light microscope.
>
> We've had "bad" gene microscopes for about fifty years. What's happened? Forensics has been revolutionized. Genealogy has been revolutionized. What's happened in medicine? Al-

> most nothing! Herceptin, Gleevec, and a bunch of cats and dogs. That's about to change: we're about to [launch] the world's first high-quality—meaning low-cost—gene microscope. We're about to do for the disease that kills one of eight people on this planet—cancer—using this gene microscope—exactly what the light microscope did for TB. That's why I changed from being a business productivity software guy to a genomics guy. With this technology and the computing technology available to us, we're going to change the world.[40]

I'd heard the "change the world" refrain before, but it did seem that Complete had a shot at fulfilling its goals if they could deliver the goods before the company ran out of cash. In August 2009, Complete belatedly concluded a $45 million round of funding, albeit six months later than had been anticipated, and an amount overshadowed by the $68 million raised by Pacific Biosciences a few weeks earlier. "Our timing was impeccable," Reid said sarcastically. "We started our raise on the day that Lehman Brothers failed."[41] One Complete investor, Alex Barkas from Prospect Venture Partners, called the financial crisis "nuclear winter for fund-raising." But the money was better late than never, and it would allow Complete to build its commercial genome center. The center would debut in 2010, with a goal of sequencing 10,000 human genomes that year, half the original goal. Drmanac was convinced Complete would have no trouble meeting that target. Reid argued that Complete wasn't so much a services business as a data business, with the unit of data being the human genome. The model was Google, which deals in one type of data (Web documents) and produces one type of report (sorted lists), based on a search. Similarly, Complete was taking one type of data (human DNA) and using a massive Google-like processing center to produce one standardized output (a human genome report).

Reid was surely cognizant of the threat from the other up-and-coming technologies. But by the time they were available, which was estimated to be around 2012, Reid claimed that Complete would have delivered more than 100,000 genome sequences to scientists worldwide. "At that point, we'll be the industry standard, we'll be the safe buy, the trusted data format. Somebody starting up a new technology won't be important."[42]

In October 2009, Complete executives traveled to Hawaii for the annual human genetics conference. They reported that they had just completed fourteen human genomes[43] including the genomes of a family with a rare genetic disorder called Miller syndrome. The sequences revealed not one but two mutated genes in the affected children.[44] Meanwhile, the sequence done for Jonathan Cohen, a geneticist at CIT Southwestern Medical Center in Dallas, had allowed him to track down the gene behind an extreme case of familial hypercholesterolemia, a disease causing high levels of blood cholesterol, in a Romanian family. The result surprised Complete staff as they hadn't been sure whether these sequences would have the power to identify such a single gene mutation, the equivalent of a needle in the haystack. Even better, the result suggested a possible drug therapy.

The icing on the cake was Complete's first scientific paper, published in *Science*, which described the sequencing of three genomes.[45] One of the genomes belonged to the person identified as PGP1 (Personal Genome Project volunteer number 1), better known as George Church. The sequence revealed that Church was an *APOE4* carrier, which had implications for risk of Alzheimer's disease. Although the paper mentioned that the average cost of sequencing the three genomes had been around $4,400—a tenth of what Quake had reported just a few months earlier—the Church genome came in at a remarkable $1,500. Just thirty months after the first personal genome sequence had been completed, Complete had dropped the price by a factor of 1,000.

When I next saw Reid, I asked him about the declining cost of reagents for Complete's latest genomes. "The consumables are now substantially under $1,000," he said.[46]

Call it the $1,000* genome.

CHAPTER 7

My Genome and Me

On a spectacular spring day in 2008, I drove south on Route 101 from San Francisco to the headquarters of Navigenics in Redwood City to meet my very own genetics counselor—or perhaps I should say, *genomics* counselor.

I had scant interest in having my genome sequenced back then: for starters, it would have cost more than $250,000, and I'm not sure it would tell me much more than I was about to learn. For 1/100 of the price, Navigenics could provide some valuable insight into my genetic risk of developing a number of common "actionable" diseases (meaning I could do something about it if necessary). Like the other personal-genomics companies, Navigenics downplays any diagnostic implications of its service. Unlike some of its competitors, however, it made in-house certified genetic counseling an integral part of its offering. Two months earlier, I'd accepted its offer to become a beta tester for its $2,500 Health Compass. I mailed my saliva sample in and a few weeks later received an e-mail that my results were ready.

Although I had few trepidations about meeting my genetic counselor, I was a little nervous nonetheless. My family tree, with its roots on both sides in the Welsh seaside town of Aberystwyth, is littered with a fairly typical range of common ailments. In the late

1960s, before his untimely death of testicular cancer, my father, Gareth, had conducted West End productions of *Cabaret* and *Fiddler on the Roof* starring a young Chaim Topol. I have already outlived him. I have a framed caricature of my dad, commissioned by one of his long-suffering orchestras, showing a gangly pencil-stick figure, baton in hand, long sideburns accentuating his bald pate, his feet trampling over the heads of the violinists. The cartoon affectionately illustrates three of my father's traits I'm curious to explore in my genome scan: the lean physique (if that's the right word) and thinning hair, both of which I share. But I didn't inherit the trait of perfect pitch, although I can hold a tune: when I was twelve years old, I sang at Covent Garden in *Carmen* with the great tenor Placido Domingo shortly after my dad passed away. My paternal grandparents died in their sixties of cardiovascular disease. My maternal grandmother, Essie, died swiftly of bladder cancer, and my great aunt Gwyneth, whom I fondly remember teaching me how to pronounce the name of the Welsh village "Llanfairpwllgwyngyllgogerychwyrndrobwllllantysiliogogogoch," lost her mind to Alzheimer's disease before succumbing to cancer. In my last visit with her in a nursing home, not long after she'd been caught trying to jump out of a window, she did not recognize me.

Navigenics leased office space in the heart of Oracle country, a metropolis of gleaming buildings surrounded by man-made lagoons. I found its office on the fourth floor of an office building that it shared with MyTriggers.com, an Internet start-up that scoured shopping Web sites. It didn't convey the ambience I had expected for a revolutionary twenty-first-century health care company; it was cheap office space by Silicon Valley standards. I was ushered into a conference room overlooking a deep lagoon. Sunning itself on the balcony outside the window was a seagull, contemplating the prospects of plucking a juicy morsel from the watery depths below. It was a metaphor of sorts for what Navigenics was about to do for me: sift a few dozen morsels, or SNPs, of information from the depths of my genome.

Elissa Levin, Navigenics's petite director of genetic counseling, has a bright smiling face framed with long curly hair. I imagine she could take the sting out of delivering the most devastating medical news if it came to that. She is one of only about 2,000 certified genetic counselors in the United States, with a master's in health care supplemented with pre-med rotations in molecular genetics, statistics, psychology, and counseling. "I couldn't just do bench work," she told me "I really liked the human aspect of it."[1]

After training in her native Pennsylvania, Levin was hired in 2002 as a counselor at the University of California, San Francisco. "I saw everything, from families with a child with a particular syndrome to adults with cancer syndromes—the whole gamut. We had people coming from Oregon, Nevada, the Central Valley. . . . People were driving eight hours, waiting for months to visit these clinics." Seeing the need to expand genetic services, she joined a Web-based, direct-to-consumer start-up, DNA Direct. "I want to make sure these people are doing this right," she thought. She ran the firm's counseling program for three years for a dozen clinically validated genetic tests. "I'd get calls from people all the time: 'I want information about my risk for this common disease.' I kept saying, 'We're not there yet.' " But with uncanny timing, Navigenics was established precisely to provide information on risk factors for common diseases. "This is going to happen, it *is* happening," she told me. "All of our core skills translate over to what we're doing here at Navigenics."

At launch in 2008, Navigenics furnished information on eighteen common conditions, including diabetes, heart disease, obesity, Alzheimer's disease, and breast and prostate cancer. Unlike traditional genetic counseling, which focuses on mostly rare diseases caused by a single-gene (Mendelian) variation, where the consequences are typically yes or no, Levin and her two fellow counselors had to tackle common diseases where the risks are conveyed in terms of above- or below-average probabilities. These conditions are called complex diseases because the risks are influenced by a combination of genetic and environmental factors. Levin often has little idea of estimating what proportion of the total genetic risk they have uncovered for any given disease. She encounters a range of reactions: "Yeah, I've always known I was at risk for this," or "Now I know I really have to pay attention," or "Wow, my risk is low!" "It's our responsibility as genetic counselors that we're not missing something else that this test doesn't cover, because this isn't whole-genome sequencing," she said.

Levin stressed that professional counseling is an essential part of the Navigenics service. She aimed to provide a welcoming and reassuring resource to subscribers around the clock—even people just contemplating the service: "We're here if people have questions. We have specific hours when we're absolutely available, and if that doesn't work, they can call and schedule an appointment."

I asked Levin to explain the difference between a genotyping service and a diagnostic test, which Navigenics and its peers vehemently insist they are not providing. "It's a great question," she replied. "There's room for error on a DNA chip," she continued, meaning it may be appropriate to refer clients to take a more specific, certified diagnostic test. "Most of the information we're providing is predictive information. We're taking retrospective information. We know that these SNPs are significant, but we don't quite have the positive predictive value proven. When you get to some of those prospective data is when you get a diagnostic. What is the accuracy of the actual test? How confident are we that that SNP is actually representative of what's in your genome at that locus? And part of it is, with what confidence can we tell you what to do with that SNP information? I think that's really when you cross the threshold of a diagnostic."

Using an Affymetrix chip that had been designed to map variation across the genome rather than analyze specific genes of medical interest, Navigenics focused on common diseases. Even if it wanted to, it couldn't offer much in the way of information about monogenic diseases such as cystic fibrosis, muscular dystrophy, or Bloom syndrome. (Other consumer genomics firms have done a more thorough job in that arena, as we shall see.)

The conditions Navigenics selected to include in the Health

Compass were based on advice from its counselors, medical director, and a team of PhD scientists and epidemiologists that met weekly to evaluate the latest research findings. The scientists curate the latest genetics advances, and the counselors and clinicians underscore the conditions that are the most medically relevant to clients. Each disease is assessed on an "actionability scale" to decide what consumers can do with the information. For example, Navigenics did not include ALS (Lou Gehrig's disease) not because there haven't been legitimate gene associations but because it's a nontreatable condition for the time being. Including Alzheimer's disease in the Health Compass was controversial—not all experts deem it to be an actionable disease—but Navigenics cofounder Dietrich Stephan believed strongly in the value of offering consumers a "therapeutic window"[2] based on early detection of genetic risk, rather than waiting until irreversible symptoms are detected.

With that cheerful admonition, I typed in my password on Levin's laptop, and prepared to meet my genome.

This was not my first flirtation with my own genetic code. In 2005, I had contacted Sciona, a company based in Colorado in the fledgling field of nutrigenetics. The MyCellf DNA evaluation kit was designed to provide dietary and lifestyle recommendations gleaned from individual genetic data.[3] After filling in a rigorous lifestyle questionnaire, I had sent it off with a cheek swab and received the printed results a few weeks later with a sense of déjà vu. The report soberly recommended that I should exercise more, cut back on alcohol and caffeine, and eat more cruciferous vegetables.

Brilliant, I thought. *I've known that for years.*

The MyCellf test looked at specific DNA variants (or genotypes) in about twenty genes chosen for their suspected relevance to various medical conditions. Among the genes tested were those coding for bone density and diabetes, heart disease, inflammation, and high blood pressure. Still, the ensuing dietary recommendations—increase my intake of folate and omega-3 fatty acids—would be standard medical advice from any family physician. In a few cases, specific gene variants prompted more personalized dietary advice in the form of recommended vitamin and antioxidant supplements. This couldn't hurt, but would they actually do a body good? A *Newsweek* cover story on the nutrigenomics fad said it best: "Some people will be advised to eat broccoli, while others will be told to eat . . . even more broccoli."[4]

While I was digesting the results of my heart and bone health, Sciona was also targeting genes aiding sports performance, or what it branded "DNA Fitness." If that sounded far-fetched, it didn't bother the Manly Sea Eagles, a professional rugby league team in Sydney, Australia, that began genotyping its players for genes related to exercise, muscle type, and oxygen consumption, so that coaches could customize training regimens for their players.[5] (The last time I checked, the Eagles had just won Australia's rugby championship, so maybe it was working.) But the field was ahead of its time and suffered a deathly blow when an official government report blasted some companies for hawking bogus dietary supplements. Another start-up company, NutraGenomics, summed it up: "Eat less, exercise more, and choose your grandparents wisely."

I also had tried the IBM Genographic Project, a collaboration between IBM and the National Geographic Society.[6] I mailed in my cheek swab, which for $99 would be checked for a dozen or so variants on my Y chromosome. Because Y chromosomes are by definition passed from father to son, this would give a sense of my patrilineal history. The results showed, not surprisingly, a path out of Africa, through the Middle East, turning left into Western Europe. More important than the specifics, however, was the demonstration that hundreds of thousands of people were willing to spend $100 and send their DNA through the U.S. mail.

The welcome page of the Navigenics Health Compass causes Levin to blush at the embarrassing picture (later removed) of her introducing a video tutorial. The results are not displayed straightaway, which sensibly gives users the option of selecting which conditions they want to see. In this way, a client can bypass a particular condition, as James Watson did with Alzheimer's disease. "Do you mind me looking at your results with you?" Levin asked, as she customarily does with her clients. "What are the questions you have up front?"

Some people, she explains, have specific questions. Others want to delve into genotypes and odds ratios, which calculate the relevance of an SNP to a trait, and the population average. "What we present here," she explained, "is the estimated lifetime risk based on the average lifetime in America. What's the chance you can develop the particular condition?" For example, for heart attack, the male lifetime average is 42 percent. Most of this information initially is based on European/Caucasian populations, but where possible, information is provided for African-American or Asian populations.

The all-important results screen shows a grid of colored panels, each representing a condition, arranged in five columns from left to right according to my estimated lifetime frequency: under 1 percent, 1 to 10 percent, 10 to 20 percent, 30–40 percent, and above 50 percent on the far right. (Happily for me, that last column is empty.) An orange square signifies a 20 percent (or more) increased personal risk compared to average, or an overall risk of at least 25 percent (regardless of how one compares to the national average).

Of all the reports provided by consumer-genomics companies, I find this by far the clearest and most intuitive. Most of the panels are colored a reassuring gray, meaning my risk is no higher than average. I counted five orange panes out of seventeen, which I thought was a good start. I also know that the diseases I'm most likely to contract are skewed to the right of the screen, regardless of whether I'm above or below average risk. Levin explained: "If there's a really common condition and you have a slightly increased risk, as you might for heart attack, you might not take that as seriously as a more uncommon condition, where you have a thirty-, forty-, fifty-fold [increased] risk. So this allows people to think about those conditions."

One of the orange boxes is labeled obesity, which made me

chuckle. The Health Compass put my lifetime risk of obesity at 31 percent—right in line with the national average of 34 percent—and thus alerts me with an orange box. Levin sensed why I was laughing. At six feet two inches and 165 pounds, my problem is gaining weight, not losing it. As a teenager, my mother would beg me to drink Mackeson Irish stout in a vain attempt to pack on some pounds. On the next screen, under the heading "What's next?" I read: "Painful question: How much did you weigh in high school?"

Clearly this isn't the whole story. A Web graphic estimates that 67 percent of the risk of obesity is determined by genes rather than environmental factors. But Navigenics screened for variants in only two genes, and you don't need a PhD in molecular genetics to suspect that many more genes must be involved in shaping body weight. Because we carry two copies of every gene, my genetic risk is calculated from a maximum possible four risk markers. I carry just one risk variant—in the "fatso" gene, officially known as *FTO*.[7] The Navigenics site confirmed that "researchers conservatively estimate that 100 to 200 other genes and their interactions, as well as their interactions with lifestyle factors like diet and exercise, may be implicated in obesity." If obesity is two-thirds genetic, then I must have a lot of low-risk variants lurking in my sequence, presumably counteracting *Fatso*. Levin sounded a little bashful. "There's a lot more to come. These numbers could change over time." Indeed, I'd wager a few pints of Guinness on it.[8]

Another orange panel on my Health Compass chart flags prostate cancer: Navigenics reported an average lifetime risk of 17 percent, but my result was 24 percent. This left me a bit puzzled: Should I be alarmed at a 40 percent increased risk above the population, or put it down to "noise" and margin of error? "Was this a condition you were concerned about?" Levin asked. In short, no. Most prostate cancer risk is attributed to the environment. I have no family history of prostate cancer, and my PSA level, which Levin says is the best tool we have, is normal. I noted to myself that the four listed risk genes were originally published in my former journal, *Nature Genetics*. "We're looking at four different SNPs, and we know that there are many more to come," Levin continued. "It's a really interesting region," she continued, spoken like a true geneticist, referring to the spot on chromosome 8 where three of the four risk factors congregate. "It's a very gene-poor area; they don't have any idea what its significance is in terms of functionality." She added that this was the best estimate right now. "Other SNPs will come along. The fact you don't have a family history is reassuring." Her bottom line: "It's a fine line between evidence-based medicine and preliminary data." I mentally circle prostate cancer for follow-up.

We briefly talked about a few other conditions. There is reassuring news for me for Crohn's disease, multiple sclerosis, and type 2 diabetes. I have a slightly elevated risk of the curiously named restless leg syndrome. "It wasn't a health condition before there was a treatment for it," Levin said, but now there's a syndrome, a treatment, and a risk assessment. My heart attack risk coincides with the national average. "Heart disease is the progression; heart attack is the outcome. There are a lot of other SNPs currently in curation," she said.

At last we turned to Alzheimer's disease. My eyes have been shifting over to the box, which, of all the initial results for various complex diseases, is the most definitive—and potentially, I thought, the most depressing. The gene in question is apolipoprotein E (*APOE*)—the one Jim Watson had redacted from his sequence. In the early 1990s, Allen Roses and colleagues at Duke University had uncovered a robust association between *APOE* on chromosome 19 and late-onset Alzheimer's disease. The association appears to account for as much as two-thirds of the hereditary risk of the disease. In my case, the Alzheimer's disease panel has a reassuring gray background: my risk is half the population average. "The odds ratios we're looking at for many of these conditions are 1–2, whereas the Alzheimer's variants confer a four- to eighteen-fold increase," Levin explained. "So proportionately, it gives it more weight."

Throughout our session, the results for each condition were accompanied by recommended actions to mitigate that risk. These are typically commonsense suggestions—improve diet, increase exercise—that hardly justify a $2,500 investment. It begs the question: Isn't this all terribly premature? It's not just reactionary medics who are in a lather about "recreational genomics." I recalled a quote in the *Wall Street Journal* article announcing Navigenics: Johns Hopkins medical geneticist David Valle, director of the McKusick-Nathans Institute of Genetic Medicine and one of the most respected geneticists in the country, believed "progress in this field is extremely encouraging, but it's premature for this kind of test to be offered. It is too early for the tests to have any meaningful impact on the current practice of medicine."[9]

Levin nimbly sidestepped the question: "We have the standard, 'Eat better, exercise.' Some of us need a little extra push. . . . We're looking to engage in research to see if people are making behavioral modifications on this data. Yes, people are actually taking this to heart and making some of those changes." For example, people who have a high risk for celiac disease may seek further medical treatment, such as a gluten challenge. For colorectal cancer, Levin pointed out that 75 percent of people never get a colonoscopy. She noted, "That might be the motivation, the impetus you need to follow medical guidelines. We come up with all kinds of rational excuses: 'Oh, it's not in my family, or I live a healthy lifestyle.' It's giving you an insight that you wouldn't otherwise know."

Navigenics aimed to engage the medical community too. "This is not a direct-to-consumer-only interface," said Levin. "We want to improve people's health outcomes. The last thing we want is to give them this information and the doctor says, 'This is useless, go away.' " But Navigenics faces an uphill battle to drag the average family physician into the twenty-first century. Levin's one-word impression of the current level of physicians' genetics education? "Dismal."

I asked Levin what I should do with the 100-page medical report that I've downloaded from the Navigenics Web site and printed for the future edification of my GP. "It depends on your relationship with your doctor," said Levin. "Do you want to take the chance of this ending up in your record? These are preliminary data."

The passage of the Genetic Information Nondiscrimination Act (GINA) largely alleviated those concerns. But if, for example, an individual left a group health plan and opened a new individual insurance plan, the new insurance provider would go through his or her medical records. "There is risk there, because the current protections don't carry over to individual plans," Levin pointed out. "Insurers don't know what to do with this information."

Navigenics did not exactly encourage its clients to download their raw SNP data, which can be used in other software and tools, as we'll see shortly. I received a four-page disclaimer via e-mail that I had to sign before receiving an encrypted file containing the genotypes of my 500,000 SNPs. Among the potential concerns was that my identity could be compromised or I could become clinically depressed if I learned unwelcome news based on unauthenticated analysis of my raw data. "We don't have enough protections in place, we don't have enough confidence," said Levin. "Potentially someone could identify you based on the SNP profile."

I asked Levin if, as one of America's leading genetic counselors, she didn't feel like the proverbial kid in a candy store. "I'm one of those people who loves coming to work every day," she grinned, before rushing off to another meeting. I cast one more glance out of the window to the balcony and the sun-splashed lagoon below. The seagull had disappeared.

Some time later, I reconnected with Levin to flesh out my family's medical history, an important component to help counselors make the most of my results and decide whether further tests might be warranted. Her final words are vaguely reassuring: "Wrapping up your family history, I don't see anything remarkable."

I've also looked at what the personal-genomics companies in addition to Navigenics make of my DNA. deCODEme offered information on more than thirty conditions, as well as information on ancestry and some personal traits, including hair and eye color and, unfortunately in my case, baldness.[10] A big green check mark on the male pattern baldness page told me in no uncertain terms that, yes, I am likely to go bald before the age of forty. Not that I needed a gene test for that. More surprising, I learned I'm a carrier for hemochromatosis—a common disease among Europeans that interferes with the body's ability to break down iron and is treated by the medieval but highly effective process of bloodletting. The discovery was ironic, as I'd published the paper on the identification of that gene a decade earlier.[11] I was also given a "recreational" tool to compare my genome to dozens of people of different ethnic ancestry, including the deCODE CEO. That told me I am genetically more similar to Craig Venter than, say, a Mbuti pygmy, which was not altogether surprising. deCODE's service also provides access to its gorgeous Java-fueled genome browser, which allowed me to pinpoint all of my SNPs and match them against the reference genome, zooming into the raw sequence if desired.

When I had interviewed him back in 2008, deCODE cofounder Kari Stefansson had dismissed Navigenics and 23andMe as basically dot-com companies that were marketing the Icelanders' discoveries (deCODE scientists had conducted much of the original research identifying the SNPs used in all of the consumer genomics tests).[12] On one occasion, deCODE published two SNPs associated with skin cancer (basal cell carcinoma), immediately introducing those SNPs into the deCODEme service, and within twenty-four hours, 23andMe had alerted its clients that they could search for the same SNPs in their raw data. So although Stefansson downplayed the others' offerings, the truth is that once discoveries are published, all of the services are free to include them.

I found that 23andMe provided the deepest dive into the genome, presenting information on ancestry, traits (earwax, heroin addiction), carrier status for Mendelian disorders, and risks for common diseases. The research reports 23andMe provides assign one to four stars for about 100 conditions, based on 23andMe's independent assessment of the original research. Heroin addiction, for example, merits two stars. 23andMe's ancestry feature also goes far beyond mere mitochondrial and Y-chromosome testing; the ancestry painting function uses an algorithm to compare groups of SNPs from all twenty-three pairs of chromosomes against those cataloged in the HapMap project. Other services had said my genome contained traces of African or Asian genetic material, but the 23andMe report determined that my chromosomes are 100 percent European.

As others have noted,[13] the consumer-genomics companies do not always agree on their results for the same disease. Take type 2 diabetes. All the three companies discussed so far predict I have a below-average risk, but they differ in what they think my personal risk is. 23andMe puts my lifetime risk at 14 percent compared to the average of 24 to 25 percent, but Navigenics says my risk is 17 percent, and deCODEme calculates my risk at just 11 percent. Should I have more faith in the deCODE result because they looked at variations in twenty-one genes, whereas 23andMe included only nine? I honestly don't know.

Another discrepancy occurs in my results for prostate cancer. Based on five genes, 23andMe says my risk is 26 percent compared to the average risk of 18 percent in reasonable agreement with Navigenics, which puts those figures at 24 percent and 17 percent, respectively. But whereas deCODEme initially put my risk as above average as well, the subsequent incorporation of additional markers changed the calculation. Now, with more than twenty markers in its prostate panel, deCODEme projects my risk as 13.4 percent—below average. This drives home the point that results for any condition can change, sometimes dramatically, as additional markers are incorporated in a person's results. Most experts agree that there is still a long way to go before all of the genetic factors for these complex diseases are identified.[14] These discrepancies highlight how provisional the information is that these services are able to provide and why they do not claim to be offering actual diagnoses.

Twenty months after the debut of 23andMe and others, in 2009 San Diego–based Pathway Genomics debuted, offering arguably the most comprehensive health and ancestry test on the market for a mere $348. An engineer by training, Pathway founder and CEO James Plante had a long-standing interest in developing consumer products, and had launched several companies, including Smart-

Drive. He became more proactive about his own health after his father was diagnosed with polycystic kidney disease. When his father died from complications following a kidney transplant, Plante termed it a wake-up call. His goal was to to give consumers access to their genetic information at low cost. "I can go to the Life Extension Foundation and order any blood test I want online—which I do. Any diagnostic information I can find a place to get it. But for genetic information, that didn't seem to be the case."[15] In 2008, Plante met David Becker, who had previously directed the search for risk genes for Alzheimer's disease at TorreyPines Therapeutics. Plante had little difficulty persuading investors to fund Pathway and wasn't worried that he was late to the consumer-genomics party. He doubted that eighteen months had been enough time for his competitors to gain a meaningful market share. "The market for this is *everyone* if it's acceptable and the price is right," he said. With its own 10,000-square-foot clinically certified lab, Pathway has the advantage that it can process DNA samples quickly and securely on its own premises, not needing to ship samples to a contract organization.

Chris d'Eon, vice president of marketing, said Pathway had "better science, deeper results, and a better price" than its rivals. "We're going deeper than anyone else," he claimed. "We're going *one level* deeper! If you're [mitochondrial haplotype] J2A1 in 23andMe, we're going to go one level deeper, into J2A1A."

Overseeing Pathway's public relations strategy was a successful West Coast advertising agency, Spark PR, cofounded by former Netscape executive Chris Hempel who, like Plante, had her own deeply personal reasons for empowering consumers with genetic information. Hempel's twin daughters suffered from a rare, recessively inherited disorder that occurs more frequently in Ashkenazi Jews: Niemann-Pick Type C, sometimes known as "childhood Alzheimer's disease."[16] A failure to metabolize cholesterol results in deposits building up in tissues and the brain. After Hempel's daughters began to exhibit the symptoms, including lack of balance and cognitive difficulties, she received nationwide publicity for her courageous efforts to identify a new therapy for the disease: a simple sugar-like molecule called cyclodextrin that had won FDA approval to give the experimental drug to her daughters.[17] "That was the reason she was interested in coming out of semiretirement and personally taking on another new client—because this was so personal for her," Plante told me. And whatever tricks Spark had up its sleeve, one thing was sure: "We're certainly not going to start launching Zeppelins," Plante said, a dig at a notorious 23andMe publicity stunt.

I took the Pathway test as well, receiving information on more than ninety complex diseases, carrier status for rare diseases, and drug response markers. Not all of those traits and disorders would be "actionable." "I don't want to 'diss' our competition, but that's an overused statement," said Becker.[18] "If you're reporting on Alzheimer's disease, that's not really a condition you can do too much about." But Becker felt it was okay to trust people with their genetic information and motivate them, if necessary, to take action. Similar to Navigenics, Pathway featured a full-time team of genetic counselors on staff, although there was an extra fee for this because Pathway's main service was so inexpensive. Becker said eventually he'd love to give customers their full genome sequence. He'd revised his own lifestyle in light of his own genetic data, but he was trying to keep his genetic results in perspective. "I have a 100 percent chance of dying," he said. "I'm not really going to fret over a few percent increase in risk over one thing or another."

When my Pathway results were in, I called Emily Enns, one of the certified genetic counselors on staff, to walk me through them. Emily had been with Pathway only a few months and had conferred with clients for just three weeks. She was learning almost as much as I was. The results were not presented with relative risk figures but in color-coded tiers apparently modeled after the Homeland Security threat levels. There are five levels: green ("Live a healthy lifestyle") means below risk, and beige ("Learn more") connotes average population risk. Diseases tagged yellow ("Be proactive") or orange ("Take action") are ones for which the analysis shows risk beyond the population average, and red ("Immediate attention") is reserved for known mutations, such as Mendelian diseases.

My Pathway results for health conditions did not offer anything beyond the other direct-to-consumer (DTC) platforms. There were no conditions in the two highest-risk tiers, and only four out of twenty-four in the slightly elevated (yellow) risk category, including coronary artery disease and peripheral artery disease. My prostate results were in the beige (average) category, which was interesting in light of some of the other results. Pathway chose not to reveal the calculated absolute risk, perhaps because of the preliminary nature of much of this analysis. Pathway's communications director Maurissa Bornstein told me that the company chose not to include an aggregate percentage risk for any given disease in its reports "because there are other factors that may contribute to the development of disease, such as environment and lifestyle, and we feel that presenting a specific number implies a degree of precision that is simply not there. Instead, Pathway reports the genetic, lifestyle, and population information in a side-by-side model so that the consumer has a comprehensive set of data points and is empowered to review all of this information and discusses next steps with their physician."[19]

The carrier status and drug response insights were, as advertised, a definite strength of the Pathway presentation. Of the thirty-seven conditions for which Pathway tests for carrier status, I was positive for two: hemochromatosis, which I already knew about, and galactosemia, a rare inborn error of metabolism that is one of the conditions tested in newborn screening. I was negative for cystic fibrosis (CF) and was impressed that Pathway tests for no fewer than seventy-nine documented mutations in the *CF* gene; 23andMe tests for less than half that number. Pathway also looks for many of the most common diseases on the Ashkenazi Jewish panel, including Bloom syndrome, Canavan disease, and Tay-Sachs disease. In addition Pathway tested for nine drug response traits, including drugs for cancer, seizures, and HIV. I carry two copies of the wild-type *CYP1A2*1* allele, which makes me a fast metabolizer of caffeine, and based on variants in two genes, I'm not especially sensitive to warfarin, an anticoagulant drug. I would be likely to have a typical response to Plavix and a decreased response to tamoxifen, a drug used in treating breast cancer. Ironically, Pathway didn't tell me if I'm at risk for breast cancer, because those results are given only to women.

Another interesting service that offers yet a deeper dive into your gene pool than the personal-genomics companies typically deem appropriate is Michael Cariaso's Promethease program. The thirty-something Cariaso is a software engineer based in the Netherlands who is glued to his MacBook like an umbilical cord. One evening, he apologized for cutting our conversation short, but he'd just remembered he hadn't eaten in twenty-four hours. Cariaso developed a wiki called SNPedia with his business partner, biologist Greg Lennon, offering an encyclopedic collection of information on all medically relevant SNPs. Just as Wikipedia is a community resource of information on the Web, SNPedia allows anyone to post medically relevant data on any DNA variant. "SNPedia, in and of itself, is a useful resource for the planet," Cariaso said.[20] "The fact that I wrote a tool that knows how to get information out of there better than anyone else is good enough for me." By the summer of 2010, SNPedia contained information on nearly 12,000 SNPs for which there was some phenotypic information.[21]

Several years ago, Cariaso and Lennon formed River Road, a small contract research outfit intent on studying two important biological specimens—their own. Cariaso's first task was to find someone to draw a blood sample. He drove around NIH and the Washington, DC, suburbs in vain, until he saw a group of firefighters and paramedics relaxing outside a fire station. "Look, this is really difficult to explain," Cariaso said sheepishly. "It's like a little science project. I just need some blood drawn." The men exchanged nervous glances, but someone eventually agreed to draw his blood. Cariaso was ushered into the back of an ambulance, but just as the needle entered his arm, "God help me," he remembered, "the fire engine alarm goes off, the whole station empties out." Cariaso leaped out of the ambulance with just enough blood in the tube. From there, he obtained his genotypes from Affymetrix, which is what led to SNPedia.

Cariaso's latest creation is a free downloadable program called Promethease. It analyzes the list of hundreds of thousands of SNPs produced by any personal-genomics company or genome sequence and compares the individual's SNP data to the information in SNPedia. The result is a customized report of all identified and putative disease and trait associations—a much longer, and in many cases more speculative, assessment of one's genetic proclivities and predispositions. Unlike the consumer-genomics outfits, Cariaso doesn't vet the SNPs or screen the validity of the information (this is a community wiki after all). What he does do is subjectively assign each association an "importance" score from 1 to 10, with 10 being the *APOE4*-Alzheimer's risk. Among his own results, Cariaso learned he had a fourfold increased risk of glaucoma, a reduced ability to metabolize warfarin, and an increased risk of rheumatoid arthritis. "I've always had joints that pop and crack," he noted.

Cariaso vividly demonstrated the potential of Promethease when he won a 23andMe–sponsored competition to predict some of the traits and characteristics of "Lilly Mendel," an anonymous demo account holder at 23andMe's Web site, based solely on her raw SNP data.[22] Cariaso concluded that Lilly Mendel was a Caucasian with "beautiful blue eyes, brunette hair . . . and an interesting personality." One SNP was linked with higher scores on anxiety-related personality traits: "She's a night owl who likes to finish her evening with french vanilla ice cream and a glass of white wine," he wrote. (SNPs showed she could digest milk and taste bitter.) Her jogging habit might have some adverse health events, given her propensity for hypertension, heart attack, and stroke. Her doctors put her on statins, with her genes showing a good response to the drug. And he warned of an increased risk for glaucoma and macular degeneration.

The identity of Lilly Mendel was eventually revealed to be none other than company co-founder Linda Avey herself.[23] She told me that Cariaso's predictions were uncannily accurate. She's an avid jogger and tends to be anxious, she admitted: "My dad was that way, my grandmother too. It's the Norwegian side, my stomach always in a knot!" But one prediction was a little off target. Pointing to her Scandinavian blonde tresses, Avey laughed at Cariaso's conclusion that she was a brunette. "This is my hair color since I was a child, just so you know!"

With Cariaso's help, I've been able to take the SNP data from three scans (deCODEme, 23andMe, and Navigenics) and run them together on Promethease. The result comes in the form of an HTML file (a long Web page), a laundry list of my variant SNPs—at last count, nearly 6,500—with information provided by SNPedia. The first group consists of SNPs that are unusually rare or that Cariaso deems of particular interest. I already knew what the first SNP was going to be, as Cariaso had given me a preview. "Oh my," he said, looking at my results on his laptop and sounding genuinely impressed. "You have not one but *two* baldness genes."[24]

The report makes for fascinating reading, at least for me. In addition to risks I've already learned about, I carry an SNP in the "bad driving gene," though I would dispute that assessment. I have the *CYP2D6*10* variant, which makes me a poor metabolizer of medicines such as codeine, dextromethorphan (found in Nyquil), beta blockers, and antidepressants. (This I share with Jim Watson, who used the information to reduce his regular dose of beta blockers to control his blood pressure.)[25] There's an SNP that prevents weight gain on rich diets and a height gene. And, interestingly, I have one homozygous variant (rs4474514) that bequeaths a fourfold increased risk of testicular cancer. I also have information on my ability to metabolize nearly 100 medicines. Finally, Cariaso informed me that I was the first G;G homozygote they'd found at SNP rs53576 in *OXTR*, better known as the empathy gene! Based on psychological evaluations of 192 Berkeley students, the authors concluded that those with G;G homozygotes are less stressed, less likely to startle when blasted by a loud noise, mellower, and more attuned to reading the emotions of others.[26] (My full Promethease report is publicly available at www.snpedia.com.)[27]

Besides the commercial genotype services I've discussed in this chapter, it is also worth mentioning the Coriell Institute's Personalized Medicine Collaborative, a free, nonprofit study to determine the long-term impact of personal genomics by enrolling 10,000 volunteers. The offering is modest compared to the commercial outfits, but it is a valuable service for people who might not want to part with $500 or $1,000 but do want to participate in a useful research study.

According to CEO Michael Christman, volunteers who find they possess two copies of a risk variant for, say, cardiovascular disease take the news "very seriously" and make major changes to their diet and exercise. That happens much less often when people learn they carry just one faulty copy, or even if they have a known family history of the disease. "That's a misunderstanding on their part, but genetic information is perceived as special by people," he said.[28] The Coriell project is trying to inform consumers about the new genetics, clarifying issues such as the risk distinction between complex traits and Mendelian disorders, or the prospects of correcting errant genes with gene therapy (a belief held by 11 percent of respondents according to one survey). The educational mission extends to physicians as well. Christman said his project is starting small, limited to potentially actionable conditions as chosen by an all-star scientific panel. It's a pragmatic approach, intended to reassure rather than scare off medical professionals with millions of data points.

A few weeks after my genome counseling session at Navigenics, I visited my GP for my annual physical. I left my Promethease printout at home but decided to bring my personal 100-page Navigenics report. Alas, my physician had little interest in what Navigenics said about my health and wellness: "The only test I might order from time to time is the Alzheimer's test," he said. The others he reckons are too preliminary.

I marked the prostate cancer page and pointed out that my risk was above average. I also relayed the story of deCODE cofounder Jeff Gulcher's prostate, but still my GP wasn't impressed. "Grade 6 prostate cancers are typically the least serious cancers we see," he said. Despite the wonders of modern technology, he insisted on the same tried-and-trusted digital exam that David Agus joked about. As I assumed the required position, I heard the familiar snap of a rubber glove. Time seemed to stop for a while, but no more than ten seconds later, my GP gave me the news. "Small, supple, and symmetrical," he said. Now what sort of genome scan could beat that?!

The last condition I was particularly curious about was perfect, or absolute, pitch, which probably contributed to my father's musical talent. I knew better than to expect to find a gene associated with this condition, but that may soon change. For more than a decade, University of California, San Francisco researcher Jane Gitschier has been conducting a dedicated effort to map a gene that, she believes, in conjunction with exposure to early musical training, can give rise to absolute pitch. In 2009, Gitschier's group mapped such a gene to chromosome 8.[29] When the gene is eventually found, it will give researchers a new way to explore the effects of nature and nurture on a remarkable human behavioral gift.

And although the environment is not the main subject for this book, let's not forget its role. A few years ago, my recreational Sunday morning soccer days came to an abrupt end when my face intercepted a thunderous shot at point-blank range. The force of the impact had me literally seeing stars. The result was a partially detached retina that no genome scan could have prepared me for. My ophthalmologist said that I had a common condition called lattice thinning, likely in part hereditary, which would make me more susceptible to a retinal tear. I looked up lattice thinning on 23andMe, but was disappointed to find nothing there. Yet.

CHAPTER 8

Consumer Reports

Do you want to know the future? How long you're going to live? How—and even when—you're going to die?

Veteran television news anchor Liz Hayes's melodramatic voice-over introduced a segment on Australia's version of *60 Minutes* entitled "The Killer in You."[1] Hayes and her producers had persuaded a couple of Aussie celebrities—"Queen of the Waves" former world surfing champion Layne Beachley and Scott Cam, cohost of the *Domestic Blitz* TV show—to volunteer for a simple DNA test and share their results.

A glimpse of the killer within. Knowledge that could help save your life.

Both stars had reason to be curious about their genetic legacy. Beachley was adopted, a date rape baby who knew very little about her family history other than some incidence of glaucoma on her biological mother's side. She knew nothing about her father.

She's bravely handing over her DNA for a test that will give her a glimpse into her future.

Cam's parents had developed cancer, his dad dying at age fifty-four. As Cam mailed off his cheek swab, he took an admirably philosophical approach to his impending results: "If they're good, we party. If they're bad, we party!" The testing was performed by deCODE Genetics, but to my amusement, the *60 Minutes* producers invited Bob Williamson, my former PhD supervisor, to deliver the results. Through a combination of good genes and a regimental squash schedule, Williamson looked as youthful as when he had publicly berated me in the hallway of St. Mary's Hospital for sequencing artifacts twenty-five years earlier. He had emigrated to Melbourne from London a few years after narrowly failing to wrest the biggest prize of his career, the cystic fibrosis gene.

"Hit me with it, Bob," said Cam cheerily a couple of months later, when the results were in. A bespectacled Williamson peered at the deCODE printouts before informing Cam that he had double the population risk for Alzheimer's disease. That wasn't exactly good news, but Cam was far more relieved that he had no signs of any unusual genetic risk of cancer. He summed it up succinctly: "So what we've worked out is: Stay off the smokes, go for a run round the block, do some crosswords, and I *can* have a drink. I love your work, Bob!"

After learning that she too had an increased risk for Alzheimer's disease, Beachley said she'd exercise her brain by playing sudoku or learning another language. Flashing the competitive spirit that propelled her to seven surfing world championships, she said, "I refuse to take this test home and go, 'Okay this is what I'm in for.' It's still about your health and the environment and maintaining a healthy body and a healthy mind."

Williamson then cast an admiring glance at Hayes, the immaculate blonde anchor who had also done the deCODEme test. In what might have been a Freudian slip, he said, "You actually have a very nice set of genetic figures here."

Despite excellent anecdotal evidence of the medical and psychological power of consumer genomics, it is just that: anecdotal. Although the number of tangible medical success stories, from celebrities to biotech founders to the man on the street, continues to grow, they represent a small fraction of the tens of thousands of genotyping tests performed since late 2007. 23andMe's Linda Avey admitted to PBS Television's Charlie Rose that some clients are disappointed when they receive their results. "They were thinking they were going to see something far more dramatic," she said.[2] Many others are nonplussed, unimpressed, or bemused. The results of these tests resemble a TV news weather report when the meteorologist offers a 60 percent chance of precipitation, when what viewers really want to know is simply: Do I need an umbrella tomorrow? Most geneticists, let alone the lay public, are going to have a hard time interpreting the news that their lifetime risk of type 2 diabetes is 27 percent instead of the population average of 22 percent, or that their obesity odds ratio is 1-to-3.

Since consumer genomics arrived in November 2007, many scientists, journalists, and other personal-genomics pioneers have shared their results, their reactions, even their fears. Their experiences offer a picture of what genome scans can reveal—and just how much has still to be understood.

Journalist Thomas Goetz got the ball rolling with a cover story in *Wired* introducing 23andMe, happy that he did not have the genetic susceptibility to heart disease that afflicted his paternal relatives, but he worried about an increased risk of glaucoma.[3] In the United Kingdom, Mark Henderson, a *Times* (London) science correspondent, embarked on his test with some trepidation, but with the newspaper writing the $1,000 check, he became eager to participate in the genomic age.[4] After taking the deCODEme test, Henderson said his most dramatic findings were a 2.4-fold increased risk of glaucoma, which was not surprising given some family history, and celiac disease, which was. He then convened a quartet of geneticists to look over his report and relayed their criticisms to deCODE's founder, Kari Stefansson, himself a neurologist and a physician. A major concern was the prospect of false reassurance because the tests did not take into account Henderson's family history. What if someone was told she had a low risk for cancer because the test did not include the specific mutated gene in her family? Nor did the test communicate the change of risk with age. For example, Henderson was told he had a lifetime risk of asthma of 18 percent, but asthma is much more common in younger people; now in his thirties, he was symptom free.

Stefansson thought both points were fair and vowed to address them, but he curtly dismissed two other criticisms. One was from an expert who worried about the lack of counseling and raised the prospect of "real psychological distress" without it. Stefansson deemed that patronizing and both resented and rejected the criticism. As for concerns that physicians would have trouble interpreting the gray area of probabilistic information, he fired back: "I'm not going to be held responsible for the poor education of GPs. This helps to empower the patient."

Another journalist quick to take the plunge was Boonsri Dickinson, an assistant editor at *Discover* magazine who was then in her twenties. She took all three consumer-genomics tests—deCODEme, 23andMe, and Navigenics—and found interpretation was hampered by her mixed Asian-European heritage.[5] (Most genetic and clinical studies are done with European populations, so there are more data to assess individual risks for Europeans than other population groups.) Based on risk markers for atrial fibrillation (heart palpitations that could lead to stroke), Stefansson advised her to limit alcohol and coffee. She also learned, not to any great surprise, that she carried markers for Crohn's disease (she admitted she had a sensitive stomach) and macular degeneration (poor eyesight). Asked if her lifestyle had changed since undergoing the tests, she said, "I wear sunglasses now and I eat spinach."

One of the most impressive deCODEme stories (besides that of Jeff Gulcher, which I discussed earlier) belonged to a fifty-five-year-old Texan who was tested at the urging of his physician because of a history of heart problems—and discovered he had a doubled risk of prostate cancer.[6] A biopsy turned out positive, but doctors felt they'd caught the tumor sufficiently early for effective treatment.

deCODE also tested the first lady of Iceland, Dorrit Mousaieff, who proved to have a below-average risk for almost every condition on their list. "When it comes to diseases, it breaks my heart to have to tell you you're terribly uninteresting. . . . You have one amazing genetic profile," said a smitten Stefansson. "This is pretty much the genetic profile you would pay for, for your child."

23andMe commandeered the lion's share of the consumer-genomics media coverage, which found a sexy Silicon Valley start-up with Google in its DNA more appealing than a struggling Icelandic biotech. *New York Times* Pulitzer Prize–winning reporter Amy Harmon accepted the firm's offer to be one of the first guests to try "Googling" her DNA.[7] "There are two kinds of people in the world: those who want to know whatever there is to know about themselves, no matter how unsettling—and those who don't," she reflected later. While Harmon found a surprising number of people who belonged to the latter camp, she had a "masochistic preference to confront the risk-benefit analysis encoded in [her] own DNA," like catching a glimpse of her reflection. Her dislike of milk since childhood was digitally encapsulated in the presence of the lactose intolerance gene. (Actually, it was a nonmutated gene: the ability to digest milk came about due to a mutation that spread thousands of years ago.) She also had lower risks of type 2 diabetes and multiple sclerosis, and only a mildly increased risk of Crohn's disease. Harmon joked, "I was in remarkably good genetic health, and I hadn't even been to the gym in months!" But she drew the line at testing her three-year-old daughter, on the grounds that she did not want her daughter's life to seem predestined. "If she wants to play the piano, who cares if she lacks perfect pitch? If she wants to run the 100-meter dash, who cares if she lacks the sprinting gene?" Harmon could be forgiven for taking some journalistic license, but for the trait to manifest itself, musical training must play a role. And a gene that is associated with fast-twitch muscle fibers does not necessarily a sprinter make.

On Valentine's Day 2008, ABC's *Nightline* anchor Martin Bashir, best known for his infamous Michael Jackson interview, broadcast his experience with 23andMe.[8] "When did you realize that saliva could become a business opportunity?" he asked a bemused Linda Avey. Avey was on hand as a palpably nervous Bashir saw his results for the first time. Myocardial infarction: "I'm average—fantastic!" he exhaled. Multiple sclerosis, reduced risk: "This is a great day!" But type 2 diabetes was a near-doubled risk: not such a great day. Bashir vowed to lose weight and tackle his genetic predisposition to obesity.

Journalist Cassandra Jardine learned that she had a one in two chance of passing on a baldness gene to her sons, and, more important, a sharply increased risk of macular degeneration and glaucoma. She called the exercise "riveting."[9]

The brain trust at 23andMe did a magnificent job in eulogizing the benefits of consumer genomics, but even investor and director Esther Dyson admitted its limitations: "It's fascinating. But *medically* useful? No."[10] Avey told Charlie Rose about one 23andMe customer who found he had a higher risk of celiac disease, but the result really helped his daughter, who had been misdiagnosed with pancreatitis. After a further workup and a change of diet, she felt 100 percent better. Anne Wojcicki told Rose that one of the most practical uses of 23andMe was to learn if you carried certain rare genetic variants that carried a very high risk of a certain disease. As an example, she gave a mutation in one of the cascade of blood clotting factors, factor V Leiden, named after the university where the mutation was discovered. The mutation, carried in 3 to 5 percent of people, markedly increases the chances of blood clots and thrombosis or embolism, and is particularly relevant for women taking birth control pills or frequent fliers cramped in economy. There followed a priceless exchange between the billionaire and the talk show host:[11]

Wojcicki: I imagine you probably travel a lot, and you may or may not sit in economy class depending on the budgets these days, and if you are . . .

Rose: When was the last time you rode economy class?!

Wojcicki: All right, all right. Touché!

Avey: Ask *me* that question!

23andMe also received an endorsement of sorts from Dr. Mehmet Oz, then the resident medical expert on the *Oprah* show. Oz tried the 23andMe kit when Wojcicki and Avey appeared on the show in 2008. He said he hoped that consumer-genomics tests would facilitate a more personalized approach to prescribing drugs and that people would embrace the opportunity to react to genomic insights by improving their diet or make more informed choices about prescription drugs. When it came to his own genome, Oz declared that he would counter his predisposition for macular degeneration by eating leafy green vegetables: "Since I don't want to be a blind surgeon, I know I can do things today that will change my life down the road."[12] The 23andMe results also indicated that Oz had a 30 percent below-average chance of developing prostate cancer. "Thanks to this test, I don't have to have rectal exams," Oz concluded cavalierly.

While Oz may have meant he didn't need to schedule exams quite so frequently, it is this sort of false reassurance that has doctors and medical geneticists worried about the potentially adverse impact of direct-to-consumer genomics. Oz's relief failed to recognize that he still has a sizable finite risk of prostate cancer, or that the story is not over. Additional markers will assuredly be identified, necessitating that his (and everybody else's) risk be reevaluated. At that point, Oz might find his risk increased.

One of the problems with the early days of consumer genomics, which I touched on in the previous chapter, is that the results do not always agree from platform to platform and company to company. David Duncan noted this in his book *Experimental Man*, as did NIH director Francis Collins in *The Language of Life*.[13] Collins underwent all three consumer-genomics tests under an alias. "Maybe I'd get a slightly different kind of customer service than the average person," he suggested.[14] Collins also encountered some discrepancies in the results. He was told his risk for prostate cancer was increased, reduced, or average depending on which platform did the testing.[15] If the self-described field marshal of the Human Genome Project can't divine the right answer from such conflicting results, what hope is there for anybody else? Are the companies testing all the markers that have been validated? How are they computing the baseline disease odds ratio? Collins called it "a vexing problem," with a danger that the public will lose confidence in the validity of the field if the companies, all staffed with hugely talented geneticists, biostatisticians, and software engineers, can't agree on disease definitions and risk calculations. And when SNP chips are superseded by whole-genome sequencing, the problem won't go away—it will likely get worse.

Others too have highlighted the issue of platform-to-platform variability. Geoffrey Shmigelsky, a Canadian software developer specializing in iPhone apps, blogged his comparison of the three original consumer-genomics platforms.[16] Although there was widespread agreement, he found conflicting results for several conditions, including rheumatoid arthritis, ranging from a relative risk of 0.7 (Navigenics) to 1.5 (23andMe), with deCODE at 1.0. He also found disagreement in warfarin sensitivity, which could have medical ramifications.

Perhaps the most powerful evidence came from members of Craig Venter's research institute. Writing in *Nature*, Pauline Ng and colleagues compared the results from 23andMe and Navigenics for thirteen diseases and five anonymous individuals.[17] Fully one-third of the results did not agree qualitatively. For six of the thirteen diseases, the two companies provided opposing results for at least one individual: one platform declared an increased relative risk, and the other concluded the risk was below average. In one glaring example, 23andMe presented a fourfold increased risk of psoriasis in one person, whereas Navigenics found only a 25 percent increased risk. Ng attributed that 400 percent difference to one marker that 23andMe included but Navigenics did not in their respective calculations. Such inconsistencies resulted from the fact that companies used different DNA array platforms and that the respective teams of scientific curators did not always agree about the validity of certain markers.

As Ng, Venter, and colleagues also pointed out, while the DTC companies do an excellent job of estimating the genetic versus environmental contribution to each of the diseases, they did not calculate or reveal what proportion of the genetic variation their markers tested for. It's common sense that someone's risk of obesity cannot be computed by the identity of one base out of 3.1 billion, but does that variant contribute 5 percent of the genetic variation or 50 percent? Among the Venter group's other recommendations was to directly test medically relevant markers (rather than rely on tagged proxy SNPs) and place more emphasis on pharmacogenetic markers that can assess sensitivity to various drugs. Sequencing would provide a comprehensive catalog of DNA variants, but that still left the daunting challenge of understanding their biological impact. "However," said Ng, "accurate and complete reporting is a necessary predecessor to a precise functional understanding of genomic data for the consumer."

The niggling inconsistency among the consumer-genomics companies is a potential problem for personal genomics. These companies are a league apart from many of the snake oil companies seeking to cash in on ridiculous tests by playing on parental pride or lay public gullibility. But if the leading scientists in the field can't reach consensus on calculating an individual's absolute risk of common disease, what is the perplexed consumer supposed to think? Linda Avey acknowledged this problem. "It's not that these three companies intended to create this confusion," she blogged, but efforts to try to forge industry standards in 2009, mediated by a nonprofit group called the Personalized Medicine Coalition, didn't take root.[18] "The worry is that the medical community will understandably throw up its hands when confronted with these confusing and conflicting risk profiles. When it comes to translating research information into clinically actionable decisions, this is *not* the desired outcome." One idea was to set up a Web-based clearinghouse for genetic association discoveries that 23 et al. could use to calculate clients' risk assessments.

Another issue, raised earlier, is that although genes are constant, the results can—and do—change over time. This is entirely to be expected, as new markers are discovered, validated, and incorporated into the analysis for each trait. However, the consumer-genomics companies should make it clearer that the results they present are merely a snapshot in time, and are almost certainly bound to change, for better or worse, as more gene associations come to light. In 2009, *New Scientist* commissioned a pair of Dutch epidemiologists to study this phenomenon using computer simulations.[19] They produced 1,000 simulated populations of people with different SNP combinations related to type 2 diabetes. They then calculated how the predicted risk would change as new SNPs were validated and incorporated into the test. After deCODE updated its type 2 diabetes panel twice in 2008, the Dutch team calculated that one in five clients would have been reclassified as above-average risk to below average, or vice versa.[20]

If you Google ABC Television's Terry Moran, the top-rated video is the *Nightline* anchor's flub at the 2008 Democratic convention in which he described Barack Obama as "the son of a black man from Kenya and a white man from Kansas. No matter what your politics, that is a moment for the history books." A year later, Moran decided to delve into his own family history following his mother's death from Alzheimer's disease, or what his friend Meryl Comer called "the dark side of longevity."[21] Herself a former television anchor, Comer had cared for her demented husband for fifteen years and was now dealing with her mother's diagnosis of Alzheimer's disease. As president of the Geoffrey Beene Alzheimer's Foundation, Comer organized a photo shoot for GQ magazine featuring the "rock stars of science" posing with pop stars such as Sheryl Crow and Will.I.Am of the Black Eyed Peas. Among the science superstars were former NIH director Harold Varmus, several leading Alzheimer's researchers, and David Agus, the cofounder of Navigenics, clad in an incongruous black leather shirt while mugging for the camera with Seal.

The Alzheimer's *APOE* test is one of the most daunting and

clinically relevant gene tests. Journalist Anna Gosline learned she had an above-average risk of Alzheimer's disease, which had afflicted her grandmother. "Maybe I should sit back and chill," she wrote. "But when scary orange boxes glare at you from your computer screen, it's hard. You just want to do something."[22] Initially the *APOE* association with the disease, made by Allen Roses and colleagues at Duke University in 1992, made little sense biologically and even less genetically. But it has stood the test of time. The *APOE* gene comes in three flavors: E2, E3, and E4. One in four people carry one copy of the E4 allele and have a two- to threefold relative risk of the disease. One in fifty possess two copies of homozygotes of E4—in geneticists' jargon—and have around a fifteenfold greater risk. There is a ten- to fifteen-year difference between the average onset of Alzheimer's in *APOE* E4/E4 individuals and *APOE* E3/E3 individuals.[23] As Jim Watson illustrated in 2008, *APOE* is a powerful piece of information, which some people choose not to know.

Accompanied by his wife, Moran nervously traveled to the offices of Navigenics, where he sat across from chief counselor Elissa Levin to learn his results.[24] Levin told Moran his lifetime risk for Alzheimer's disease was indeed above average at 19 percent, but it was empowering news. "Could have been worse," he exhaled. Comer was not so lucky. "Oh, my God," she said as she scanned the printout of her genetic scan, which showed a 37 percent lifetime risk for the disease. "I will put my life in order," she said, dabbing away tears. "I will tell everybody how much I love them. I will live the moment. It's a piece of information, and I'll fight harder."

That sense of empowerment is a recurring theme in consumer genomics, and one that should help reassure the medical establishment, which has fretted that people would be unable to handle potentially negative genetic information. At a conference in Boston in 2008, the editor-in-chief of the *New England Journal of Medicine* (*NEJM*), Jeffrey Drazen, a fierce skeptic of the predictive power of personal genomics, outlined what he needed to see from the genetics community: "I'm from Missouri, and you have to show me . . .

that making a difference in [genetic] knowledge will make a difference in how people behave."[25] Sitting in the audience, Boston University neurologist Robert Green seized the microphone, and told Drazen his manuscript was in the mail.

Back in 2001, Green had launched the REVEAL (Risk Evaluation and Education for Alzheimer's Disease) study to determine the responses of people with a family history of the disease when they learned their own genetic risk. *APOE*, Green says, is the only gene for a common disease that has a relative risk ratio over 2 and "the only one that meaningfully increases someone's risk in a way that could conceivably mean something to an individual." Colleagues repeatedly told him *not* to embark on REVEAL, because of major concerns over discrimination (there were no federal safeguards in 2001), psychological harm, and the lack of any real treatment. But what if, as some trials suggested, the efficacy or side effects associated with trying to treat Alzheimer's depends on knowing one's E4 status? "It seemed a little patronizing, even in 2000, to say, you know what? We can give you that information, but we're not going to."[26]

Recruiting his subjects by making cold calls, Green enrolled 162 family members of Alzheimer's disease patients contacted to sign up. The benefits, he told them, included helping research, preparing the family, and arranging personal affairs, adding that there was *nothing*—not exercise, not vitamins, not drugs—they could do to change their risk. That was a blessing of sorts: it meant studying risk in absence of market pressures. And unlike Huntington's disease, a dominant inherited neurodegenerative disorder where one faulty gene is a virtual death sentence, a lot of people wanted to know this information. Surprisingly, Green found few, if any, differences in terms of anxiety, depression, or distress between those who were told their *APOE* genotype and the control group, even one year after the study. There was a transient increase in anxiety after receiving results, but it soon waned. Nobody felt like jumping off a roof, he said. The individuals who showed the most clinically meaningful changes in psychological profile were spread evenly between the control group and the disclosure group (regardless of E4 status), and while subjects moved up and down on the anxiety scale, such shifts occurred regardless of whether they had tested positive. "As long as people understand we're dealing with a different kind of probability risk, it will have nowhere near the same emotional impact" as learning a black-or-white test result such as Huntington's. Green concluded that REVEAL supported the safety of disclosing data regarding genetic-counseling protocols to Alzheimer's family members, "despite the frightening nature of the disease and the fact that the disclosure has no clear medical benefit," and thus it contradicted the prevailing paternalistic attitude in the medical community. But Green admitted the results could have been worse if the subjects had been prone to anxiety or depression or were less educated. "It's like a new drug," he said. "Is this information safe for people to have sitting in the back of their mind for ten years?" How would a subject feel if she or he forgot a name or saw a close relative succumb to the disease?

Drazen published Green's findings in the *Journal*,[27] but perhaps the most dramatic result was not included: people who learned they were E4 positive were three to five times more likely to purchase long-term-care insurance. Green presented that observation in front of a "wild and crazy group" of long-term-care insurance underwriters. "This quiet, docile, quantitatively oriented group virtually stormed the stage when they heard about this—because this kind of information will put them out of business." They are thinking very hard about whether they can offer this product without requiring E4 testing.

Of all the testimonials of consumer genomics, one might have supposed that the distinguished leaders of the nation's largest genome centers would be among the most fervent embracers of personal genomics. Curiously that wasn't the case. The Broad Institute's Eric Lander, who almost single-handedly wrote the classic report of the first draft of the human genome,[28] was in no hurry to read his own genome. "I find it very anxiety provoking—I don't want to know any of this about me," he said in 2008. The only exception was Tay-Sachs disease, because as an Ashkenazi Jew and a father, that is highly relevant information for him. During a town hall meeting at the University of Washington, Lander said:

> I have not had my genome sequenced. I have no burning desire to get my genome sequenced, because there's no particular thing I want to know for which it is the answer. . . . But if, God forbid, I had cancer, I'd get my cancer sequenced in a heartbeat and compare it to my normal genome to look at whatever I could learn, because there is something I could imagine doing with it. So for me it's very much tied to being the answer to a relevant question.[29]

Lander's colleague, physician David Altshuler, was one of the chief architects of the HapMap Project and the 1000 Genomes project. But he too sees little use for his personal genome: "From a clinical point of view, it's just noise. No one knows how to use such information to improve health."[30] Altshuler raised several hypothetical concerns. What about a woman with a family history of breast cancer who learns she has a lower-than-average risk of the disease based on a handful of SNPs, and failed to appreciate that the test does not include the classic *BRCA1* breast cancer gene? Or a man relieved that his genetic risk of diabetes is no higher than average, and so starts eating a richer diet?

Lander's counterpart at the Baylor College of Medicine Genome Center, Richard Gibbs, the Aussie who led the team that tried to interpret the first personal genome, was excited to read the highlights of his own genome but not terribly impressed with what came back. Gibbs submitted his spit sample to a Houston biotech company he cofounded, SeqWright, which had quietly launched its own personal-genomics service as a side business. His first public verdict on the exercise was blunt. "It's bloody useless!" he said, by which I assumed he meant that he hadn't discovered any major disease risk factors. Over a drink, I suggested to him that that wasn't necessarily a bad outcome. Gibbs admitted he hadn't thought of it like that, and

I mentally patted myself on the back, thinking I'd made him change his mind. A few weeks later, it was clear I hadn't.

Speaking at a conference in San Diego, Gibbs said consumer genomics was experiencing the same sort of hype that had marked the Human Genome Project (HGP), at least until some real data started to hit the floor. "If you're not sure how much hype there is, how much reality there is, I suggest you go out and get yourself genotyped," he said. Gibbs called the exercise a "visceral experience," but in the end, he said, "it reminded me a lot of those Jiffy Lube $39.95 brake check-ups." His report said, for example, that his risk of macular degeneration was increased from 1 in 5,000 to 1.8 in 5,000—not a big deal, he said, because he is myopic and has light-colored eyes. With a glint in those eyes, he said: "Kevin Davies told me that's a very cynical view, to say it's completely useless, because in fact I learned about many things I didn't have. I assume he meant things like trisomy 18!"[31] referring sarcastically to a genetic disorder caused by having an extra chromosome.

One leading geneticist who did subscribe to the "no news is good news" philosophy was Harvard Medical School professor George Church, who has done more than almost anyone else to make personal genomics a reality.

It is hard to miss the lanky 6'4" frame of George McDonald Church in person. Nor is it difficult to find details of his medical history, for it is all publicly available online: eye color (green) and blood type (O+). Medical history (narcolepsy, hyperlipidemia). Drugs (statin, omega fatty acids, and vitamin supplements).[32] Oh, and an exhaustive list of nearly 1,000 gene variants and their medical relevance, just the first steps in the latest, some might say craziest, plan in Church's illustrious career developing technologies for reading and writing DNA. The idea behind the Personal Genome Project (PGP)—sequence the DNA of 100,000 volunteers cheaply and post the results online, along with personal and medical details—would have been ridiculed coming from just about anyone else.

To look at Church, it is difficult to believe his father, the late Stew "Barefoot" McDonald, was a member of the Water Ski Hall of Fame (almost as implausible as the fact there *is* such an institution!).[33] Church, who was later adopted, became obsessed with science and computers at Phillips Andover Academy and Duke University, but in 1976, he was kicked out of Duke's graduate program for not attending classes. (He proudly displays the official dismissal letter from Dean Katzenmeyer on his Web site.) He joined Wally Gilbert's lab at Harvard, and from his early DNA sequencing experiments, when the cost of sequencing a single base was about $10,[34] schemed about large-scale personal-genome sequencing. "Six billion base pairs for six billion people had a nice ring to it," he said.[35]

Church earned a reputation as a leader in DNA sequencing technologies even before the HGP. In 1994, he helped a Boston biotech complete the first bacterial genome sequence.[36] Later, his team developed one of the first next-generation sequencing methods, called polony sequencing.[37] Elements of this technology were incorporated into the technologies developed by ABI and Complete Genomics, while the Church lab helped build a low-cost open-source sequencing instrument called the Polonator. Today he opens his lectures with a humorous conflict-of-interest slide, which is a de facto list of virtually every major next-gen sequencing and consumer-genomics company, including Knome, Helicos, Complete Genomics, 23andMe, and at least a dozen others.

But Church wasn't just interested in developing research tools. He stated the case for personal genomics before just about anyone else.[38] "I would pay $10,000 to get my genome sequenced rather than buying a second car," he said back in 2001. He believed that people would need their personal genome as much as they needed a personal computer and would pay closer attention to their health if they had such information. If people were crazy enough to pay

$35,000 to clone their cat, there was surely a market for personal genomes at $10,000 and falling. People could save lots of money over their lifetime in early diagnoses and preemptive medical care with smart nutritional and lifestyle decisions. "I wouldn't mind having my blood sequenced for prostate cancer every day," Church said.

Church also championed more openness and less stigmatization in genomic medicine. The passage of GINA (the Genetic Information Nondiscrimination Act) notwithstanding, in the long run, genetic discrimination would be pointless because every individual's genome will reveal vulnerability to some health problem. There were widespread concerns about how individual genetic information might be stolen or misused by insurers, employers, the police, neighbors, commercial interests, or criminals.[39] There were few safeguards to genetic privacy, vividly illustrated a few years ago when a teenage boy ingeniously used a bit of do-it-yourself genetic sleuthing to trace his sperm donor biological father.[40]

In 2006, Church boldly launched the nonprofit PGP, laying out his goals in an article for *Scientific American* entitled "Genomes for All."[41] He would recruit nine volunteers plus himself, who would put all their medical and genomic data into the public domain and "begin exploring the potential risks and rewards of living in an age of personal genomics." He continued: "Just as personal digital technologies have caused economic, social and scientific revolutions unimagined when we had our first few computers, we must expect and prepare for similar changes as we move forward from our first few genomes."[42]

The PGP would gain strength as more and more volunteers were sequenced, building an invaluable resource for investigators to study complex diseases and foster a new age of personalized medicine. The utility of the first personal genome was analogous to the first fax machine or Web page—the true value wouldn't emerge until communities built up around the early adopters. Church praised the PGP-10 pioneers as "heroes and human guinea-pigs paving the way for potentially increasing utility for the general public," putting themselves and their families at risk because of the long-term rewards for society.

After a year-long debate with Harvard's Institutional Review Board (IRB), Church was finally given the green light. Among those agreeing to join Church, who was PGP-1, were health care investor Esther Dyson; Harvard Medical School chief information officer John Halamka; Helicos cofounder Stanley Lapidus; Duke University geneticist and author Misha Angrist; and Harvard psychologist and author Steven Pinker.[43] Dyson penned an op-ed in the *Wall Street Journal* explaining her decision to volunteer for the PGP-10:

> I want to show that there's nothing especially magical about my genome. It doesn't hold secret knowledge that will allow others to harm me as they might by sticking pins into a voodoo doll. In fact, I feel more trepidation about releasing other information that is more personal because it will reflect my behavior (yes, I inhaled, but not a lot!), the fact that I saw a shrink (but not for long), and so on. I don't have any deep secrets or vulnerabilities that would embarrass or create risks for myself, or for relatives who share my genes. I don't have an employer who could fire me for black marks on my health record, and I have health insurance . . . which will cover me for whatever ensues from the precancerous colon polyp that was discovered and removed a couple of years ago.[44]

Lapidus volunteered not so much because he was curious about his genome but because he wanted to join the debate about genomic privacy. Approaching sixty years old, he was in reasonable health but admitted, "I will surely die! The right to know your genome and right to privacy should be viewed as human rights, but I'm interested in exploring (with me as an experimental subject) the opposite."[45] On balance, he placed a higher priority on keeping his financial data private than his genome.

Church had already set the stage by posting his medical records online, and he had been rewarded. During a lecture at the University of Washington in 2004, a hematologist in the audience pointed out that Church's cholesterol level was far above the normal level of 200. Church went back to his doctor, who doubled his prescribed dose of lovastatin. "In the future, this kind of experience would not rely on transcontinental serendipity but could spawn a new industry of third-party genomic software tools," Church predicted.

In summer 2007, the ten PGP pioneers donated blood, skin biopsies, and cheek swabs, providing not just DNA but establishing cell lines. As it was still not economically feasible to sequence ten whole genomes at $100,000 a pop, Church chose to focus on just the genes, the 1 to 2 percent of the genome (known in the trade as the exome) that code for proteins, where most of the medically relevant variants reside. (The rest could wait until next-gen sequencing technologies got faster and cheaper.) In November 2008, the PGP 9—Dyson was in Russia training to become a cosmonaut—held a press conference at Harvard to discuss their early data. A local TV station branded the launch of Church's "genetics Facebook" as "one small online posting, one giant leap forward for medicine." The only PGP pioneer who was initially hesitant to post his entire genomic data online was Angrist, who, as the father of two daughters and with a family history of breast cancer, wanted to review his sequence data privately before releasing them.

PGP-2, self-described "geek doctor" John Halamka, learned he had a doubled lifetime risk for coronary heart disease, which didn't surprise him. He had packed on 100 pounds in his twenties, the result of getting married and "the CIO all-stress diet"—too many Starbucks lattes and supersize Big Macs.[46] Even though he had become a vegan (shedding all the excess weight), he wished he'd had the sequence information in his twenties. He also carried a rare mutation for hereditary motor sensory neuropathy, a condition that manifests with muscle weakness and numbness. Interestingly, his father suffered from multiple sclerosis, raising the possibility that the two were connected. His father volunteered to join the PGP to answer that question.

Like the other PGP volunteers, Halamka had had to consider the potential risks of making his genome public. How will this information be put into a patient's electronic medical record? How is it presented to doctors in a way that is accessible and actionable? And then there was the question of how relatives, friends, colleagues, and neighbors might react to one's public information, even though Halamka had few concerns about any lasting damage to his reputation. Nor was his teenage daughter unduly concerned about any dates looking for undesirable genes in her dad's DNA.

Steve Pinker confessed to a modicum of narcissistic pleasure in poring over his early results, but he found that some of the sequence annotations in the databases left a little to be desired. "I have some susceptibility to irregular menstrual periods," he joked, as well as "a tendency to be born prematurely." He later wrote, "I soon realized that I was using my knowledge of myself to make sense of the genetic readout, not the other way around."[47] Why, for example, did he carry variants that predicted red, thinning hair, when in fact he sports an enviable mane of curly silver-black hair? Why did he like brussels sprouts while carrying a bitter-tasting gene? Pinker likened such deterministic thinking to trying to understand the stock market by studying a single trader or a movie by putting a DVD under a microscope. "The fallacy is not in thinking that the entire genome matters," he wrote, "but in thinking that an individual gene will matter."

Nevertheless, after consulting with Boston University neurologist Robert Green, Pinker said he would likely follow Jim Watson's lead and apply a "line-item veto" of his *APOE* status, even though he was reasonably confident that he could handle an *APOE4* positive result. But he wasn't going to lose any sleep over a higher risk of type 2 diabetes. Whether the risk is one in five (the population risk) or one in four, as in Pinker's case, a sensible diet and exercise regimen made sense in either case.

One of Church's hopes for the PGP was that it would spur the development of software to help physicians scan a patient's genome and identify the alarm signals, for example, to order a blood test or a colonoscopy ahead of schedule if that's what the sequence suggested. Xiaodi Wu, a student in Church's biophysics 101 class, spent his final two years at Harvard writing an open-source program called Trait-o-matic, which is being used for the PGP analysis. Trait-o-matic compares the list of personal genome variants with various data sources, with a particular focus on rare variants, though it was a daunting challenge to highlight the medically important results from a list of 1,000 or more potentially significant variants.[48]

Curiously, Church wasn't as introspective about analyzing his own genome as one would have expected. "The way I look at it is, we go down the [PGP] list, and the computer prioritizes which are the most interesting alleles that we're turning up. The most interesting allele was in PGP-6, not in PGP-1." PGP-6 happened to be Steve Pinker. When Wu reviewed a list of Pinker's gene variants with PGP medical director Joe Thakuria, they saw near the top of the list this entry:

chr12:109841347	**C/T**	**CARDIOMYOPATHY, HYPERTROPHIC, MID-LEFT**
MYL2, A13T	**T**	**VENTRICULAR CHAMBER TYPE 2**

The rare mutation (which the Church lab confirmed by resequencing)[49] was in a gene coding for myosin-light chain 2 (*MYL2*). Other mutations in this muscle protein have been associated with hypertrophic cardiomyopathy—a dangerous enlargement of the heart that can cause sudden cardiac arrest, notably in athletes such as basketball stars Hank Gathers and the Boston Celtics' Reggie Lewis.[50] After some counseling, Pinker decided to get a full checkup from a colleague, the renowned cardiologist Christine "Kricket" Seidman at Brigham and Women's Hospital.

Pinker's gene variant sat close to a known pathogenic mutation associated with sudden death. "Luckily, we have seen this identical codon 13 *MYL2* variant a few times in healthy people of the same ethnicity, so it was unlikely to be pathogenic," she told me.[51]

Even though Pinker was hospitalized once for chest pains, his EKG, echo, and other tests came back normal, again supporting the notion that this is a harmless variant. Still, she advised Pinker that he should get cardiac screening every couple of years, probably for the rest of his life.

Pinker's experience highlights a major problem for personal genomics: interpreting the medical significance of gene variants for which there is no family history. At last count, there are more than 1,500 mutated genes deemed predictive and actionable, and a huge fraction of those variants occur without family history. Figuring out whether variants in each of these known Mendelian genes is a potentially damaging mutation or simply a benign polymorphism will be a huge challenge. Moreover, the impact on health care costs for follow-up exams in cases like Pinker's was not trivial. "Imagine costs if clinical exams are needed to exclude pathogenic mutations in every Mendelian gene?" Seidman said.

Prior to his cardiomyopathy findings, Pinker published some advice for those contemplating taking a personal genome journey. Those consumed with scientific or personal curiosity and capable of thinking in probabilities should by all means enjoy the fruits of personal genomics. "But if you want to know whether you are at risk for high cholesterol, have your cholesterol measured; if you want to know whether you are good at math, take a math test. And if you really want to know yourself (and this will be the test of how much you do), consider the suggestion of François La Rochefoucauld: 'Our enemies' opinion of us comes closer to the truth than our own.' "[52]

Nevertheless, Robert Green felt the genie was out of the bottle: "I believe this trend—genome scanning or gene sequencing and its disclosure to anybody who wants it—is absolutely unstoppable." People should have access to expert consultation, he said—not necessarily their physician, because most physicians don't have much working genetics knowledge. "The question isn't really, Should it be happening, or can we stop it? The question is to foster an incremental path along which it's done better and better."[53]

CHAPTER 9

Cease and Desist

From the inception of the consumer-genomics industry, every piece of marketing collateral, Web site real estate, and legal document bearing a company logo stressed and reaffirmed the strictly educational, nondiagnostic nature of their genome-scanning services. "Information you learn from 23andMe is not designed to diagnose, prevent, or treat any condition or disease or to ascertain the state of your health," said the Bay Area company. "23andMe's services are intended for educational, informational, and research purposes only." Although Navigenics played up its on-call team of certified genetic counselors and the actionable nature of the conditions included in its Health Compass, it too stressed that its services "do not establish a doctor-patient relationship and are not intended as medical advice." The company urged potential clients "to work with your physician or other qualified health-care provider to develop an optimal personalized health management strategy." Similarly, deCODE cautioned: "The Genetic Scan product is for informational purposes only, is not medical advice, and is not a substitute for professional medical advice, genetic counseling, diagnosis, or treatment." And Pathway Genomics warned that its services "have not been fully validated and shall not be relied upon by you or any other person to diagnose, treat or prevent any disease or health condition."

Still, Linda Avey spoke for 23 et al. when she told a U.S. Department of Health and Human Services (HHS) committee that "individuals have a right to access their genetic information and learn about themselves in a new way. . . . They should not have to pay for those services through a health professional to find out those facts about themselves." And, she added, "Federal and state government physicians should not impede information development in [a] paternalistic view about what people can handle."[1] And Pathway declared: "You own all genetic information derived from your saliva."

Nevertheless, in the absence (at the time) of any federal safeguards against the sudden rise of consumer genomics, two of the most populous states, New York and California, decided to clamp down. In early 2008, New York State warned thirty-one companies that genetic tests could not be performed on samples from its residents without formal approval. Ann Willey, director of the state's Office of Laboratory Policy and by definition the enforcer of New York State's rigid laboratory testing policy, is both a board-certified geneticist and a lawyer. Her agenda was to find an appropriate way to regulate the fledgling industry. "I think of this genomic profiling paradigm . . . as really a star," she said with a matronly air. "By the time we get done regulating it . . . we're going to have to force it into a globe and shear off some of its sparkling and promising aspects."[2]

Since 1964, New York State's regulations have covered laboratories that handle any specimen derived from the human body, which covers paternity testing, forensic profiling, insurance underwriting, and employment testing. Many of the state's rules are more stringent than federal Clinical Laboratory Improvement Amendments (CLIA), and require both labs that have state permits and out-of-state labs receiving specimens from New York State to meet the same standards. Having concluded that none of the "hobby genetics" companies, as she called them, held such permits, Willey dispatched letters stating in effect, "not in New York State unless you have a permit." The ruling put a damper on the April 2008 Navigenics launch party in Manhattan, as staff cautioned that New York residents ordering a spit kit would not get their results until the state gave its approval.

Willey wanted to be flexible, but her hands were tied. "We're all in this together," she said. "We've got to come out of this the other end with the application to a new paradigm to personalized medicine." In Willey's view, the regulatory system was not singling out genetic testing companies—the safeguards applied to all forms of clinical testing. However, the emergence of personal genomics had raised some difficult questions. Even if the personal-genomics companies did not receive a body specimen directly or perform experimental analysis themselves, they typically combined that information, performed the data interpretation, and delivered the results to the client. Willey had to regulate these companies, and the difficulty was pigeonholing them under rules laid down decades earlier. Were they involved in the practice of medicine, in a laboratory, or in information management? "Once we make it a duck, it better quack like a duck," she said. "No matter what box we put it in, we put constraints on it. But we don't want to leave them in no box, because we have no oversight." Willey insisted she wasn't picking on the consumer-genomics industry. "We really want to make this work. But I'm from the government and I can't always help."

In New York, Willey's hands were tied. But no one expected the State of California to clamp down as well. Nevertheless, on June 9, 2008, Karen Nickel, chief of Laboratory Field Services (LFS) for California's Department of Public Health (CDPH), hastily dispatched certified cease-and-desist letters to thirteen genetic testing companies.[3] The warning letters, on CDPH letterhead and bearing the logo of California Governor Arnold Schwarzenegger, were mailed the same week that Schwarzenegger delivered a keynote address to a major biotech convention in San Diego hailing the state's biotech pedigree. Citing California's $73 billion in annual biotech revenue ("that's without the sales of Botox to Joan Rivers") and its passage of proposition 71 pledging $3 billion for stem cell research,

Schwarzenegger trumpeted the Golden State as "the biotech capital of the world . . . and one of the best places to set up shop." The twin towers of the consumer-genomics industry, 23andMe and Navigenics, had cause to dispute that.

Nickel's actions were prompted by consumer complaints about the price and quality of tests offered online. "People are very concerned about the accuracy of the tests and the aggressiveness of some of the companies," said a CDPH spokesperson. Cost wasn't the issue for the department, but accuracy and licensure were. "Do they even have a license? Should they be offering these tests to California residents?"[4] Nickel's "notice to cease and desist performing genetic testing without licensure or physician order" gave the thirteen offending companies two weeks to comply with other instructions or potentially face sanctions including fines up to $10,000 a day. CDPH made three key points: any business offering genetic tests to Californians had to be licensed as a clinical laboratory in the state, it needed a CLIA certificate for laboratory testing, and all genetic tests must be ordered by a licensed physician. (The providers also needed to show that the tests were validated and that customers had access to test counseling.)

Aside from the big three personal-genomics companies, other recipients included nutrigenomics firms Sciona and Suracell and genetic testing firms such as HairDX, Gene Essence, and DNA Traits. One person who welcomed the intervention was Navigenics cofounder David Agus. "I thought it was the greatest thing in the world," he said.[5] "All these mom-and-pop people were trying to get into it." Without a barrier to entry, Agus feared a repeat of the nutrigenomics scare of 2006 (a number of companies tried to link DNA variants to dietary advice). Indeed, some companies, including Sciona, immediately ceased accepting orders from California residents. SeqWright did the same, even though it wasn't on the CDPH list. Others, such as California genetic testing company DNA Direct, escaped the clampdown.[6] CEO Ryan Phelan argued that DNA Direct "set the industry standard for responsible delivery of genetic testing services," exceeding guidelines released by the National Society of Genetic Counselors and the American College of Medical Genetics (ACMG). It provided pretest consultation, informed consent, and test authorization by a medical geneticist. It ran its tests in CLIA-certified labs and produced individualized reports tailored for the patient's physician.

The major sticking point was the requirement to use a physician to order the tests. deCODE and Navigenics were founded by physicians, but did their involvement in devising, processing, and interpreting the DNA scans constitute "ordering" by consumers? Did the personal-genome services constitute a "genetic test" as described by the CDPH regulations? And perhaps most important, what role, if any, should a state have in controlling access to an individual's DNA sequence and the information encoded therein? But consumers were bypassing their physicians and ordering services directly over the Internet, and that appeared to contravene California code. "It's insulting and a curtailment of my rights to put a gatekeeper between me and my DNA," blogged California-based *Wired* executive editor and author Thomas Goetz. "This is *my* data, not a doctor's. . . . Regulation should protect me from bodily harm and injury, not from information that's mine to begin with."[7]

23andMe and Navigenics carried on business as usual while making conciliatory noises about the need for regulation and cooperation with the CDPH. 23andMe said it was eager to collaborate "in the development of an appropriate regulatory framework for the personal genomic industry."[8] Navigenics insisted it was already compliant with California law: its tests were performed in a CLIA-certified lab in California and approved by a California-licensed physician.[9]

Of the three conditions laid out by the CDPH, CLIA registration was the least onerous. 23 et al. used (or had switched to) CLIA-certified labs for genotyping, and all were taking steps to meet the state licensing requirements. Kari Stefansson, however, was irked by the CDPH's erroneous allegation that deCODE was directly marketing to California residents. "This letter was shot from the hip by babbling idiots. We've been trying to apply for a license in the state of California since the middle of [2007] and they don't respond to our emails or our letters. And then they have the audacity to say that we are marketing things without a license?!" He called the fuss over the cease-and-desist letters "a temporary misunderstanding on behalf of the bureaucracy to try to prevent direct marketing to consumers."[10] He had no problem with state or national authorities regulating how direct-to-consumer (DTC) tests were performed; indeed, he welcomed the imposition of quality control standards. But threatening to prevent consumers from gaining direct access to their personal genetic information was seriously offline, he argued. These same states allowed marketing of cigarettes and alcohol directly to people, although both have proven health risks. "But there is absolutely no study, not even indirect evidence, that it is bad for people to learn their risk of disease. All preventive health care is based on the concept of raising concerns about your diseases and about the possibility of translating that concern into preventative measures."

As for California's insistence that the tests be ordered by a state-licensed physician, Stefansson said most of deCODE's diagnostic tests *were* prescribed by physicians rather than the DTC route. Nevertheless, it was a mistake to interpose a physician on someone's ability to learn about his or her risk of disease. As examples, he pointed to the availability of supermarket blood pressure machines and over-the-counter pregnancy tests. "Do you really think that the pregnancy test has been dangerous to people, now that women are able to go to the drugstore and determine if they are pregnant or not? I think it is extremely important, it is liberating for people, and it's going to power the arrival of really valuable preventive medicine."

So would deCODE be complying with the CDPH demands? Stefansson sputtered a choice Anglo-Saxon epithet before reconsidering: "We will do whatever they want us to do. We are marketing all over the United States to physicians, so we could easily sell this test in California even if they demand it is done through physicians." deCODE opted to partner with a New York State genetic screening firm, Lenetix, that was licensed by California and New York. A short time later, 23andMe struck a similar deal with LabCorp.

Meanwhile, Navigenics brought in some legal muscle: Los Angeles health care law firm Hooper, Lundy & Bookman. In a nine-page letter, partner Bradley Tully insisted that Navigenics was already complying with the state's regulations.[11] One would have expected, Tully wrote, that the lead in the regulation of genetic testing would have been taken by the federal government, but the HHS had declined to exert jurisdiction. "Our understanding . . . is that HHS does not have sufficient expertise with respect to genetics counseling and genetic information services to regulate them, and that the purely laboratory aspects of such businesses are already adequately addressed by CLIA." Tully noted that Navigenics contracted its genotyping to a CLIA-certified and California-licensed lab operated by Affymetrix and pointed out that Navigenics was conveying information, not the lab test results, and so was not acting as a clinical laboratory. (The full list of 900,000 SNP results was transmitted only on request, with a separate consent form.) Rather, it was acting similar to a physician in interpreting lab results, except that Navigenics wasn't practicing medicine. Furthermore, if California's LFS was suggesting that Navigenics was marketing a new kind of lab test, based on a combination of Affymetrix testing and Navigenics's interpretation, Tully scotched that as well:

> All of the steps involving application of the techniques of clinical laboratory science to a biological specimen are performed by Affymetrix, which is a California licensed laboratory. Navigenics does nothing at all to "obtain scientific data." It instead interprets the data that has been obtained by Affymetrix. Nothing in the definition of a clinical laboratory test supports a conclusion that the interpretation of the data resulting from such a test is itself a test or a part of the test.

Tully also argued that LFS wouldn't object to physicians' or patients' researching information on disease-related SNPs from sources such as WebMD. Moreover, if Navigenics published its algorithms for calculating risk assessments in a white paper or an academic journal, the LFS would not contemplate jurisdiction over such sites.

The other chief complaint was that the lab tests had to be ordered by a physician, but it was already established LFS policy that "any licensed physician in California may order laboratory tests on persons of whom they have no knowledge." Tully's closing remarks were: "The tests which are performed by Affymetrix have been requisitioned by a California physician, and the reports of the tests are not reported by Affymetrix to the clients, but are instead reported to the physician who has ordered the testing."

The lobbying worked. In August 2008, the CDPH duly granted licenses first to Navigenics and shortly after to 23andMe.[12] deCODE Genetics was approved in 2009.

But after a year grappling with what label to attach to the consumer genetics companies, New York's Ann Willey reached a different decision.[13] "These entities that will obtain raw data from the analytical testing facility and generate a report that would go to the ordering practitioner are a laboratory," Willey decisively told an HHS committee in March 2009. The DTC companies were not so different from a pathologist who received slides from an analytical facility and issued a diagnosis and therefore must be licensed as a lab. "We consider these data management facilities no different," she said. "We're making these data management companies laboratories."

Willey had received a mixed response to her "Not in New York" letters. Some smaller entities did not respond, a dozen agreed not to practice in New York, and others said they would stay out of New York until they applied for a permit.[14] When potential 23andMe customers tried to order kits online to be sent to New York, they received a warning that they should mail their saliva samples from outside the state.[15] A member of the HHS committee asked Willey what would happen to those celebrity DNA samples collected by 23andMe during its New York Fashion Week party in 2009 attended by such luminaries as Rupert Murdoch and Harvey Weinstein. Willey said they had been destroyed, although Avey later told me, "It was called a spit party but realistically it was a party, with not a lot of spitting going on!"[16]

In the event that 23andMe wanted to continue recruiting New Yorkers, Willey had some constructive advice: "It's not that far to Connecticut. Have your parties somewhere else!"

Wojcicki said she was still hoping to figure out a solution with New York, because it was such a ripe constituency with the Wall Street crowd, an active Parkinson's disease community, and the media, to name a few. "It seems silly that you can just go to Hoboken [New Jersey] and have the [spit party] events there. . . . You could even push it and do it on the Circle Line!"[17]

Meanwhile, in December 2009, the State of New York issued a "clinical laboratory" permit to Navigenics, allowing it to process saliva samples originating within the state with one caveat: they had to be ordered by a patient's physician.[18]

The state government watchdogs have by no means been the only ones concerned about the sudden rise of consumer genomics. The medical establishment was swift to react, and the reaction wasn't pretty. Leading the charge was Muin Khoury of the Centers for Disease Control and Prevention (CDC) in Atlanta, but not because he's vehemently antitechnology or is desperate to deny people access to their DNA. "Anybody who wants to get their genome is fine with me," said Khoury.[19] "I love technology, but I want to see it used to help people. The hypothesis that personal genome profiles improve health and prevent disease is just that—a hypothesis. I'd like it to be tested using scientific approaches." Khoury didn't have a problem with people forking out $500 or $1,000, if they wished, but his issue was with the interpretation of that information.

Khoury came to the United States from Lebanon in 1980 to do his public health training at the CDC and stayed after the war intervened. In 1997, judging that the CDC, which is charged with protecting health and preventing disease, needed to be a leader in studying the impact of genomics on public health, he created the

National Office of Public Health Genomics. Although modestly funded, Khoury's group assumed responsibility for figuring out the "translational pathway" to turn gene discoveries into public health benefits. "Keep in mind, we're trying to reduce the burden of disease at the population level," he said. "We're trying to see how genetics fits outside the domain of genetic diseases." Although we are entering a new era of personalized medicine, Khoury argues that a population approach is needed to figure out what genes mean for people and how they interact with the environment.

Just weeks after the public launch of 23andMe, Khoury was invited by *New England Journal of Medicine* editor-in-chief Jeffrey Drazen to assess the validity of consumer-genomics start-ups. The resulting editorial bluntly warned physicians of a potential explosion in public queries about genetic testing, with the prospect of patients dropping a thick, personalized genomics dossier on their desks. Drazen's team signaled its disapproval and, by implication, that of the medical establishment.[20] Imagine, they wrote, that an overweight patient showed up in his doctor's office carrying his genotype report—a gift from his caring children. The testing had found an increased risk of both heart disease and diabetes, and the patient was anxious for some medical direction.

Khoury and Drazen argued that "such premature attempts at popularizing genetic testing seem to neglect key aspects of the established multifaceted evaluation of genetic tests for clinical applications." The claims of consumer-genomics companies, Khoury said, were based on partial and incomplete information.[21] Most adult diseases, like type 2 diabetes, cancer, or heart disease, are caused by multiple interacting genes. Measuring the association of disease-related genes one by one produced "a very incomplete picture of that complex array of associations" and minimal information on disease risk. Khoury claimed the tests had no clinical value for predicting disease and, whether positive or negative, would confuse consumers. Would most consumers even know what a "positive" result meant? Did "positive" mean a beneficial outcome, or the successful detection of a mutation? And then there were the long-standing concerns about the protection of genetic information.[22] In short, Khoury concluded, "These tests are not ready for prime time. Therefore, we should lay off them until more research."

If and when physicians were faced with patients who gave them their personal genome printout, Drazen and colleagues recommended that physicians make "a general statement about the poor sensitivity and positive predictive value of such results." And for those demanding to know whether the DTC services provided useful information for avoiding disease, the prudent answer would be: "Not now—ask again in a few years."[23]

Khoury and others had four principal areas of concern regarding the validity of consumer genomics, which they summed up with the acronym ACCE: analytical validity, clinical validity, clinical utility, and ELSI (ethical, legal, and social issues).

Analytical validity, the first concern, means whether the test can be reproduced and deliver accurate genotypes and results. "The consumer-genomics realm is really a black box," said Boston University neurologist Robert Green. "Even if the accuracy is 99.9 percent accurate, there are quite a number of errors in there."[24] The commercial microarrays from Illumina and Affymetrix have been honed over several years and are used by labs worldwide. But the Drazen editorial correctly pointed out that "even very small error rates per SNP, magnified across the genome, can result in hundreds of misclassified variants for any individual patient."

In fact, the accuracy of the DNA chips has improved significantly, pushing 99.99 percent or about 100 errors on a chip containing 1 million markers. (However, translate that error rate to an entire genome, and you'd be faced with tens or hundreds of thousands of errors.) The issue of analytical validity flared up in 2009 when a reporter for *New Scientist*, Peter Aldhous, uncovered a software glitch in the presentation of dozens of variants in his deCODEme data.[25] One genealogist who reviewed the data asked Aldhous, "Are you sure this is *Homo sapiens*?" and in June 2010, 23andMe revealed it had sent the wrong results to 96 clients because of human error (a mishandled tray of saliva samples) at its contractor, LabCorp.

If the analytical validity of each platform was beyond reproach, then one would suppose that the results of 23andMe would closely match those of Navigenics, and deCODEme would correspond with Pathway. All use established genotyping technology and sophisticated teams of scientists to run the algorithms that compute individual risk. But as we have seen, there can be surprising differences in the results between platforms stemming from their use of different sets of SNPs, baseline risk assumptions, models, and algorithms.

In an attempt to iron out those discrepancies, the Personalized Medicine Coalition hosted a series of friendly discussions with the leading DTC companies, but no agreement was reached. As Avey said, "You don't want to start driving antitrust by saying everyone should be on the Illumina platform." A wiki platform would allow more and more people to comment on associations and give the DTC companies greater credibility among the physician community.[26] Avey liked to imagine a futuristic *Star Trek* utopia where a world federation comes together to forge a common consumer genetics understanding. But that's unlikely. "You don't want to have everyone so cookie cutter that you can't set yourself apart. The one who builds the best interface and puts the most tools for the consumer world is going to win."

Clinical validity, Khoury's second area of concern, concerns the ability of the test to predict the disorder in question. *APOE* testing for Alzheimer's disease is a paradigm for consumer genomics because of its exceptional clinical validity. For now, "It is the only gene for a common disease that has a risk ratio over 2," said Green. "It is the only one that meaningfully increases your risk in a way that could conceivably mean something to an individual." One in four people carry one copy of the E4 variant, and have a threefold increased risk, or a substantive odds ratio of 3. One in 50 people harbor two copies of the E4 allele, which carries a fifteenfold increase in risk. Those odds ratios make *APOE* the most robust risk factor for developing Alzheimer's disease.

But in most instances, the relative risk attributed to a specific gene variant is much more subtle—around 1.1 or 1.2, which has limited significance. "People who are not statistically sophisticated will not necessarily understand that," Green continued. "All they'll see is that the risk is higher or lower, and they'll interpret it within a frame of reference that is far different than the statistics can support." Efforts to fully account for a person's genetic susceptibility to complex diseases require the identification of all the relevant genes and their variants for that disease, and an appreciation of how those genes interact with each other and the environment. Most of the variants offered by consumer-genomics companies were associated with relative risks of less than 1.5, which, Khoury charged, did a poor job of distinguishing people who will develop these diseases from those who will not.

Green said it was largely the consumer's responsibility to figure things out for themselves: "This is very much the entrepreneurial spirit of Silicon Valley, and in a way that's the ultimate answer. But along the way, segments of society and many potential consumers have the possibility of misinterpretation and misunderstanding." Green can't offer any concrete examples of consumer panic brought on by personal genomic testing, in contrast to the experiences of counselors and clinicians dealing with patients and families handling the emotional impact of serious monogenic disorders such as hereditary breast cancer and Huntington's disease. Consumers need to recognize they are dealing with a more subtle type of risk. So too do physicians, many of whom lacked the background in genetics or statistics to handle probabilistic DTC information. On the other hand, the overachieving, Web-savvy postgrads from elite universities running consumer-genomics companies needed to appreciate what Green tactfully called "intermediate levels of understanding among the general public."

Clinical utility, the third area of concern, is the balance of risks and benefits if a test is introduced into clinical practice. If a patient is genuinely at risk for a disease, what can be done about it? Is the test "actionable"? Navigenics claimed it considered only "actionable" diseases, but Khoury argued virtually no clinical data demonstrated the risks and benefits associated with screening for specific genes. Khoury wasn't arguing that learning about a genetic susceptibility for diabetes or heart disease was a bad thing or that it wouldn't motivate the subject to take preventive steps; rather, no tangible evidence implied that relationship. Moreover, any claim regarding clinical utility relied on the assumption that interventions that work in the general population will apply equally in a genetically at-risk population. But many of these interventions—smoking cessation, weight loss, increased exercise, and control of blood pressure—would be broadly beneficial for many conditions, regardless of a person's genetic susceptibility to a specific disease.

Khoury was also unable to give me a single story of a consumer-genomics client who had reacted adversely to test results. But he raised the case I described earlier of deCODE's Jeff Gulcher, who had prostate surgery following a positive genetic test and biopsy. "What happens if hundreds of thousands of men take that test at the population level?" posed Khoury, fearing a calamitous cascade effect. "The health care system could be weighed down under this technology, with people demanding more and more services that are more and more expensive with unknown benefits." Khoury conceded, "In Jeff's case, it may have saved his life. But it may not have." Most men die *with* a low-grade prostate cancer rather than *from* it. Khoury worried that men might turn to SNPs in calculating their prostate risk, potentially decades before the onset of symptoms, rather than rely on the conventional PSA screening, which isn't perfect but is a much more downstream marker of the disease.

Gulcher argued his surgery not only saved his life but also saved the taxpayer more than $100,000 in long-term medical care, given the likelihood his cancer would have metastasized. "Listen to me," said Khoury. "These are all testable hypotheses. This is not Jeff's word against mine." Rather, it's about doing the necessary clinical trials to evaluate the benefit of personal-genomic information.

If gene markers can distinguish those men who will get aggressive prostate cancer versus those who will be able to tolerate it and live a long life, then that, said Khoury, is a much more clinically useful tool. "If the intervention can be tailored to the personal genetic risk factors, I'm all for it," he said. In 2010, that is not yet the case.

Khoury also worried that "patients who test negative may be falsely reassured and thus less motivated to comply with preventive recommendations." He highlighted the reaction of Dr. Oz, Oprah Winfrey's medical protégé, after taking the 23andMe test, whose risk of prostate cancer was 30 percent lower than average. "Dr. Oz will become complacent," said Khoury. "You know probabilities—it doesn't mean he will never get prostate cancer. Translate that into diabetes and heart disease . . ." Testing for twenty SNPs for diabetes "doesn't tell you more or less than what you already know based on your family history and body mass index." Some people will be higher risk, some lower than average. Is the intervention the same in both cases? Shouldn't we all be exercising and eating well? Khoury insisted the onus was on DTCs to prove that their SNPs did more benefit than harm rather than the medical authorities' having to worry about potential horror stories.

The *NEJM* editorialists hoped that just as the threat of Craig Venter's Celera galvanized the completion of the Human Genome Project, the emergence of consumer-genomics firms might spark efforts to conduct the necessary translational research. But until the genome could be put to work, the children who had given their father genetic testing as a gift "would have been better off spending their money on a gym membership or a personal trainer so that their father could follow a diet and exercise regimen that we know will decrease his risk of heart disease and diabetes." Drazen, a diminutive, bow-tie wearing asthma researcher at Harvard Medical School, took the reins of the *NEJM* in 2005. A few months after his editorial tirade, he joined a roundtable discussion on personal genomics.[27] "We have to be a little careful about our ability to handle large amounts of information without making mistakes," he said, pointing on his lapel to his registration badge. "You think I'm Jeff

Drazen, but the people at the [front] desk think I'm 'Temporary Access!' " Evidently the conference computers didn't have a record of Drazen, and he was none too happy.

Drazen insisted that the onus was on consumer-genomics firms like 23andMe and Knome to prove the information they provide is beneficial, and he did not think they were close. He recalled an amusing anecdote about a newly married man who noticed that when he returned home from a road trip, he'd be really out of breath. It turned out his parents both had asthma, and that he had a family history of cat allergies. In fact, the patient had the highest antibody levels for cat allergens that Drazen had ever seen, presenting a huge risk. "I told him about this [and] explained to his wife she had to get rid of the cats. She said, 'I've been married for six months. I've had the cats for ten years.' "

Drazen felt personal genomics was far from having clinical utility. In breast cancer, the known relative risk ratios for genetic markers barely reached 1.2, and yet the lifetime risk for women in the population ranges from 6 to 16 percent—a spread of just 10 percentage points. A clinic might tell an individual woman, "We don't want to do a $500 mammogram because you have a low-risk profile." Drazen said: "You still have a risk of getting breast cancer. I mean, it's not zero."

Drazen was also nervous about data privacy, which brings us to *ELSI.* He recounted another story of e-mailing a friend to help him erect a fence at his home on Nantucket, and "the next time I sent a Gmail, someone was trying to rent me a house or build me a fence." Drazen worried that someone's personal health information would seed a new enterprise of selling services and products of dubious validity based on someone's genetic profile. Harvard Medical School chief information officer John Halamka, a member of Google Health's advisory council, assured Drazen that patient data in Google Health would never be data-mined, sold, or used for advertising. "As soon as they abrogate that trust," he said, "no one will use their services." Drazen did not look convinced.

Privacy issues were mollified by the passage of the Genetic Information Nondiscrimination Act (GINA) in 2008, signed into law by President George W. Bush. A heroine in that quest was Sharon Terry, who quit her job as a college chaplain following the diagnosis of her two children with a rare genetic disorder, pseudoxanthoma elasticum (PE), a progressive disease that produces excess calcium deposits in the body. Terry dedicated her life to understanding the genetics of the disease and bringing order to the patchwork assembly of state laws and regulations regarding the privacy of genetic information. She was a coauthor of a 2000 study that identified the gene mutated in PE.[28] Nine years later, as the president and CEO of the nonprofit Genetic Alliance, she had the ultimate satisfaction of finally seeing the U.S. Congress enact federal legislation to ban discrimination based on genetic information.

Some commentators, notably author, lawyer, and clinical geneticist Philip Reilly, say the sense of fear regarding consumer genetics is overhyped. For a couple of years, the Air Force Academy didn't consider carriers of sickle cell anemia for admission, even though they were asymptomatic, because of an erroneous concern that there would be a higher risk if a plane depressurized. But although the term *genetic discrimination* dates back to around 1986, Reilly argued that there is scant documented evidence of genetic discrimination by insurers and employers. "Whole forests have been cut down in writing the papers to discuss this, but when you actually go in our most litigious society on earth, and ask, 'Where are the lawsuits? Where is the fight?' It just isn't there."[29] He continued: "We have an adequate structure in place to protect against misuse of genetic information, but the drumbeat of concern for a whole generation preceding this new birth of consumer genomics has really created a problem in public perception that will take a long time to undo." Reilly said the harms that will occur are more likely to be dignitary than economic—hurt feelings and damage to one's self-esteem rather than a lost job or denied insurance.

But Terry insisted there have been many cases of genetic discrimination. In 2002, the Burlington Northern Santa Fe railroad agreed to pay $2.2 million to settle charges under the Americans with Disabilities Act of illegally testing workers for genetic defects. The company had performed DNA tests on thirty-six workers who complained of job-related carpal tunnel syndrome without their knowledge. Or consider the case of basketball star Eddy Curry, who balked when the Chicago Bulls insisted he take a DNA test to determine his potential risk for a sudden heart attack—a condition called hypertrophic cardiomyopathy—that had claimed the lives of several basketball stars. Curry even turned down an offer of $20 million in compensation should the test turn out positive, fearing he would be setting a dangerous precedent for his fellow athletes, not to mention potentially stigmatizing his children. Curry subsequently signed with the New York Knicks and passed his physical. No DNA test was required.

Terry documented lesser-known discrimination cases in what she called her "phone book." The children of a Kentucky woman who suffered from alpha-1 antitrypsin deficiency were denied health insurance, even though they were asymptomatic carriers of the gene. Coverage was restored after Terry brought the story to a major newspaper. The woman's daughter appealed in writing to her congressman, saying: "My mom says that everyone is created equal and deserves to be treated fairly. Please help my mom stop people from treating others unfairly." Another woman who tested for *BRCA1* didn't tell her health care provider and died of ovarian cancer because she feared using the test information. At a congressional hearing, Terry testified: "Every one of you, and each of your loved ones, is at risk for some disease or other. . . . Please remember that none of us have any choice over our ancestry, our different abilities, or our genetic makeup. As a nation, we do have a choice about how we treat that information."[30]

In 2006, IBM became the first major American corporation to outlaw genetic discrimination for its employees. Chief of privacy Harriet Pearson simply inserted the term *genetics* into IBM's existing equal opportunity and diversity policies that covered other types of discrimination based on age, gender, and other factors. Finally, by a nearly unanimous vote, the 110th Congress finally passed GINA on May 21, 2008, broadening the patchwork assembly of statewide legal protections.[31] Meanwhile, the British government extended a moratorium with the Association of British Insurers to prevent the use of genetic test results to deny people insurance until 2014. The historic passage of GINA was warmly welcomed by scientists like the Broad Institute's David Altshuler, who gave the example of the *BRCA1* gene and breast cancer. If a woman carrying a *BRCA1* mutation elected to have a prophylactic mastectomy and oophorectomy, her risk of dying would be reduced by 90 percent. "Now, people in that setting can go ahead and do what's right for their health and not have to worry about discrimination." Or maybe not. In April 2010, a Connecticut woman brought the first lawsuit under GINA when she sued her former company for firing her after she tested positive for the *BRCA2* breast cancer gene and had a voluntary double mastectomy.[32]

Protecting genetic privacy is all well and good, but a pair of *New Scientist* reporters were able to demonstrate the feasibility of hacking someone's genome. Michael Reilly swabbed a glass that Peter Aldhous had drunk from and sent the sample to deCODE to amplify the trace amounts of DNA. (Some companies like 23andMe require customers to submit saliva samples in part to minimize the possibility of identity theft, but deCODEme relies on a cheek swab.) Aldhous reported a greater than 99.995 percent match between his hacked deCODEme results and his previously obtained 23andMe data.

The *New Scientist* stunt raises the disturbing prospect of paparazzi trailing celebrities for souvenirs to air their DNA laundry. That could go as far as collecting DNA to smear presidential candidates, who are loathe to release their medical records at the best of times, so the prospect of their voluntarily releasing snippets of their genomic data seems unlikely.[33] But DNA could in principle be obtained from a loose hair, a plastic cup, or even a Rose Garden beer summit and used to imply the presence of a risk gene for Alzheimer's or Parkinson's disease or multiple sclerosis. Robert Green wrote, "In the next presidential campaign, someone might publish a candidate's genome and focus on a marker that has been linked to a psychiatric condition, regardless of how unproven the association is. . . . The threat of genetic McCarthyism provides us with an opportunity to engage in a public dialogue about the limitations and complexities of using genomic information for decisions about life and health—including voting for our president."

GINA is not a perfect law. An insurer can request genetic information, such as asking about *BRCA1* status, before covering a prophylactic mastectomy. On the employment side, GINA does not apply to military coverage or long-term disability coverage and does not cover employers with fewer than fifty employees. Finally, there are concerns that GINA could allow individuals who learn of certain genetic risks to load up on life or long-term disability insurance. Perhaps these loopholes will be closed in subsequent legislation.

In all the political wrangling over the Obama administration's health care reform bill, little attention was paid to the impact of personal genetic information. Back in 2002, William Brody, former president of Johns Hopkins University, argued in a *Wall Street Journal* op-ed that personal genomics would spell the end of private health insurance as we know it:

> In the years ahead, genetic testing will become gradually more pervasive, and at the same time, our knowledge of the risk of disease associated with the results of those tests will become increasingly refined. The result could be the end of private health insurance as we now know it. If legislatures pass laws banning insurers from using genetic screening data, those companies will protect themselves by continually raising premiums to consumers. Some may even go bankrupt because purchasers of insurance will be the more knowledgeable in the transaction. . . . Yet if we allow insurers to use genetic data, many more individuals will be left without coverage because they will be deemed too high-risk to warrant insurance at affordable prices. Given this conundrum, there is only one solution that can preserve the concept of health insurance: universal coverage.[34]

Brody's argument was not lost on 23andMe's Anne Wojcicki. "I think a future where insurers utilize genetic information in a positive way, to provide improved health care on a personalized basis at a lower per-capita cost, is within the realm of possibilities," she said. Under a single-payer system designed to encourage preventive care, "Genetic information could be used not only to screen people at high risk for disease, but to better engage individuals to actively participate in their own health care."[35]

In the wake of Drazen's strident *NEJM* editorial, several professional societies weighed in with their own policy statements on the new world of consumer genomics. The American Society of Human Genetics said, in effect, that some tests were appropriate to offer directly to consumers, but it was important they be accurate and reliable.[36] The American College of Medical Genetics (ACMG) took a sterner position, advocating that health care providers should be involved in ordering and interpreting all genetic tests, and the consumer should be fully informed about the test's validity. "This was basically putting a stake in the ground, saying we, the medical geneticist community, need to be involved in this testing," said Kathy Hudson, founding director of the Genetics and Public Policy Institute. The executive director of the ACMG, Michael Watson, said his members were the "professional guides to the human genome" and were needed to help the public make informed decisions related to genetic testing. "This is not an area where people should really 'go it alone,' " he held.[37]

Another leading medical geneticist, New York University's Harry Ostrer, typifies the old-fashioned paternalistic point-of-view with this bit of advice to would-be genomic consumers:

> Get diagnosed if you have a genetic disease. Get treated and learn about it. Knowledge is power. Learn about your family history. If, based on your family history, you're at high risk of colon cancer, heart disease, blood clots or a host of other con-

> ditions, consult with a medical geneticist—there may be a specific genetic test that is appropriate for you. Get carrier testing before marriage, before pregnancy, during pregnancy. Get the other tests that are appropriate for your stage of the life cycle. Learn your risks and act. You don't need a $1,000 genetic test to spur you into action.[38]

Ostrer looked forward to the day when a sixty-one-year-old man enters his office exhibiting telltale signs of tremor and inserts his "DNA sequence micro-disc" into the "genetic risk reader" beside his desk. Within a few seconds, the machine beeps that the gentleman has Parkinson's disease type 7 and, moreover, should be treated with a particular drug on account of his "personal risk-benefit genetic ratio." Until that time, Ostrer feared that "the purveyors of direct-to-consumer genetic testing, with their faulty claims, will scare away our patients and betray the public's trust in medical genetics."

I sought out a clearer picture of the medical establishment's attitudes to consumer genomics by calling ACMG president Bruce Korf at the University of Alabama, Birmingham. Korf wasn't so worried that the results are being handed straight to consumers without the fatherly filtering of a health professional. "At rock bottom, I'm not necessarily inclined to a paternalistic perspective that this information is too hot for the layperson to handle and they should be barred access to it," he said. And he candidly admitted that many primary care doctors wouldn't know much more about a genome scan than the curious patient would. But consumers should work with a trained geneticist because the information is still so raw and the potential to be misled is enormous.

The two primary concerns for Korf were the means of communicating the information and the quality of that information. The information conduits being fashioned by consumer-genomics companies would become robust in time, but the initial quality has been highly variable and fairly primitive. The size of the relative risks is very modest, and Korf doubted that many people could appreciate just how subtle the changes hovering around the baseline risk are. Consumers would likely drown under a ton of complex information. Even worse than having no understanding of what those results mean is having a vague idea, prompting people to embark on misguided health programs and lifestyle changes—perhaps a businessman who stops taking his statins because he learns that his genetic risk of heart disease or stroke is below that of the general population. His concerns about exercise or weight vanish—until he has a heart attack. But these are only hypothetical concerns. Korf admits he has no documented cases, although he's certain they exist.

The only solution is to do population-based studies to discern the impact of consumer genetic reports. "Let's imagine you could sequence an entire genome for a modest cost, now here you are ready to counsel them. The problem is, for the most part, you wouldn't have any good data to base counseling on. You'd have a fantastic amount of information, but not a whole lot of background about what it all means." Korf said that his patients are often still confused after an hour of counseling. Although he is skeptical now, computer programs may become available to assist or assume the burden of counseling and he would have to consider new models of counseling: "I could imagine, in decades to come, that the numbers of things people will need to be counseled could exceed the workforce we could ever anticipate having."

Publicly at least, the consumer-genomics companies have welcomed the prospect of federal regulation, not least because it would help differentiate between the companies trying to do this properly and what Robert Green calls "a raft of essentially fraudulent companies selling genetic snake oil that is being foisted on people. . . . Romantic attachments, cosmetics, nutrition, vitamins, that's where there's not a scintilla, not a scientific fig leaf. We're making a mistake focusing our energy on the leaders who are trying to do this correctly." Those sentiments were echoed by Larry Thompson, NHGRI communications director, who asked, "In the absence of regulation, who is to advise the public what is legitimate and what is little more than genetic snake oil? Could the charlatans ruin the industry for everybody else?"[39]

Some DTC companies are building their business by trading on parental fears or human gullibility. The slogan for Atlas Sports Genetics, which was not granted a certificate by the State of California, sounds like something lifted out of the movie *GATTACA*: "Finding any great Olympic champion normally takes years to determine. What if we knew a part of the answer when we were born?" The company's name stands for "Athletic Talent Laboratory Analysis System." For $149, the test examines precisely 1 base out of 3 billion in the human genome that provides "one measure of natural-born athletic ability." Based on that result, the Atlas president said parents should "nurture that potential Olympian or NFL star with careful nutrition, coaching and planning."[40] The variant in question is a common change in the gene that codes for a muscle protein called actinin-3 (ACTN3). The mutation, called R577X, substitutes a "stop" signal (X) for the amino acid arginine (R) at position 577 of the ACTN3 protein, prematurely halting production of the protein. The idea is that the presence of two copies of the R577X variant predisposes subjects to endurance, whereas the "normal," or wild-type, gene results in a muscle type better suited to sprint and power events. Elite sprinters rarely have two copies of this R577X variant, but then again, neither does some 18 percent of the global population, or about 1 billion people. When researchers in Madrid found that a Spanish Olympic athlete had two copies of R577X, they concluded there might indeed be "notable exceptions to the concept that *ACTN3* is the 'gene for speed.' "

At best, *ACTN3* accounts for 2 to 3 percent of muscle function and might have some modest relevance as to which type of sport a gifted athlete is best suited for. Daniel MacArthur, one of the Australian geneticists behind the original *ACTN3* studies, wrote: "Whatever combination [of *ACTN3* variants] your child has, he or she will share that with a large chunk of the population, the vast majority of whom will never go on to be international-level athletes."[41]

For $1,995.95, Eric Holzle's company, ScientificMatch, offers an online genetic matchmaking service that also stretches the bounds of scientific credulity. Holzle claims a host of purported benefits, including a more satisfying sex life, enhanced fertility, more orgasms (for women at least), a higher chance of monogamy (women again), and healthier children—all in all, not a bad return for a cheek swab! He got the idea watching a documentary about the classic sweaty T-shirt experiment. In 1995, Swiss biology professor Claus Wedekind asked female students to rate the odor of six sweat-stained T-shirts worn by male students. On the whole, women were most attracted to the scent of men with immune systems different from their own. Holzle's matchmaking method complements MHC genotyping with more conventional measures of compatibility. But maybe there is something in it. During a segment on Scientific-Match on CBS's *Early Show,* the host had a confession to make: "I could sniff my husband's T-shirts all day long," she said.[42]

For about $2,000, another DTC start-up, My Gene Profile, claimed to help parents reveal the inborn talents that God has given their children by revealing forty talents and personality traits in eight categories.[43] Potential customers were warned that their children could lead "a mediocre life" if they failed to use genetic testing to identify their child's inborn talents, to say nothing of their own wasted time and money.

While the U.S. Food and Drug Administration watched and waited, the Federal Trade Commission (FTC) was wrapping up its investigation of some of the early DTC nutrigenomics companies for potentially misleading marketing,[44] including Boulder, Colorado-based Sciona. British HGP leader and Nobel Prize winner Sir John Sulston said: "Nutrigenomics is a very easy scam. Not only is the advice useless . . . worse, some companies are associated with companies that will sell you the dietary supplements."[45] Sulston's advice was simply to grow your own vegetables. In the summer of 2009, the FTC suspended its investigation into Sciona,[46] but that was moot as Sciona had quietly ceased operations.

In May 2010, the FDA notified 23 et al. that it considered their kits to be medical devices "intended for use in the diagnosis of disease" and that an urgent dialog with the agency was needed. "I wonder whether 23andMe will be foolish enough to go to war with the FDA," pondered Stanford bioethicist Hank Greely. But legal expert Dan Vorhaus said the FDA flap was likely to be "a blip on the radar screen when it comes to individuals having unfettered access to all of their [genomic] information."[47]

CHAPTER 10

Another Week, Another Genome

To mark National DNA Day—the annual celebration of the publication of the double helix—in April 2009, a boutique Boston biotech, Knome (officially pronounced "gnome" except by cofounder George Church, who insisted on calling it "Know-Me"),[1] launched the first charity auction of a whole-genome sequence on eBay. With a starting bid of $68,000, the winner would be treated to a digital copy of his or her genome on a titanium flash drive and four nights at a luxury hotel in Boston while attending his or her private day-long genome consultation. Knome even threw in dinner with Church. About the only thing missing were Red Sox baseball tickets.

The auction proved a bit of a flop, attracting precisely one bid, but it was further evidence of just how fast the price of personal-genome sequencing was falling. Just two years earlier, Watson's sequence was unveiled for $1 million. And when Knome debuted six months later, the introductory price was $350,000. An 80 percent discount in eighteen months is fairly remarkable in any line of business. Unlike 23 et al., Knome's service was not some scraping of the tip of the genomic iceberg, which covered about 0.02 percent of the human genome, but rather the first commercial offering of personal, whole-genome sequencing, approaching 100 percent of the client's genetic code. That meant personalized information on thousands of common and rare diseases. The full sequence of all 20,000 genes was available at last to the general public (at least to the millionaire set), not just celebrity scientists.

"We're going to have done two of the first five genomes on the planet," said Church proudly when Knome launched. Jorge Conde, Knome's stylish young CEO, said his first twenty clients would have a historic opportunity to pioneer the emerging field of personal genomics. Conde had gotten his first taste of the genomics business after completing his MBA in a summer job at Helicos, before joining Flagship Ventures. In 2006, he organized a panel discussion on the $1,000 genome for an MIT conference with a serial IT entrepreneur named Sundar Subramanian. Afterward, Subramanian approached Conde: "I don't understand the science side of things, but if you get to the $1,000 or $10,000 genome, you have a major IT problem. What do you do with all that data?"

Conde and Subramanian then approached Church with the idea of launching a personal-genome sequencing company. Their timing was uncanny. Church was already fielding speculative inquiries from trendsetters wanting to have their DNA sequenced, but he couldn't contemplate recreational sequencing in his academic set-up. In August 2007, the start-up was incorporated as Cambridge Genomics, with Subramanian serving as chairman and Conde as CEO. His first priority was to convince Subramanian to rethink the company's name. "We laugh about this. He's had a Cambridge Devices, Cambridge Technology Partners, he has all these 'Cambridge' companies, right? I said, 'Look, walking by my apartment this morning, I passed Cambridge Dry Cleaners, Cambridge Diner . . . ' " And so Knome was born.[2]

From the outset, Conde decided that Knome would offer whole-genome sequencing, rather than a compromise "exome" approach,[3] concentrating on the 1 to 2 percent of the genome that codes for protein, which appealed to Church. "This is going to be where it ends up," said Conde, so why not start at the top and do it right? The other hallmark was a determination to safeguard the privacy and security of its clients, who would doubtless treasure their anonymity. Conde insisted he wasn't trying to avoid the Food and Drug Administration (FDA) but stressed that "this isn't a diagnostic, this isn't medical advice. This is a personal research project we're doing for individuals." If the FDA wanted to talk, he'd be happy to comply.

Knome's charter clients weren't hypochondriacs or obsessive-compulsives, but technophiles committed to biomedical research. "They're passionate and curious," said Conde. "They're fascinated with the concept of genomics, and they have the means to participate in a revolutionary technology. There's a level of altruism to what they're doing."[4] His first two clients, one Asian and one European, were "vehemently private" people who had stumbled on Knome in the press, although not the way Conde had envisioned—by spotting the tiny Knome advertisement in *Robb Report*, a magazine that typically advertises Lamborginis and Gulfstreams. One of the clients had read about Knome in *Science* magazine's 2007 "Breakthrough of the Year" issue.[5]

Knome's first European client turned out to be not so private after all. Dan Stoicescu, a retired Romanian biotech entrepreneur living in Switzerland, told the *New York Times* he would rather spend his money on his genome, which he would scan for updates like a stock portfolio, than a Bentley or a private jet.[6] There was an amusing rejoinder from Jim Watson, who was still mulling over his own DNA data: "I was in someone's Bentley once—nice car. Would I rather have my genome sequenced or have a Bentley? Uh, toss-up." Harboring some misgivings of his own, Stoicescu questioned if he shouldn't have donated his money to feeding hungry children in Africa.[7]

For $350,000, Stoicescu received Knome's luxury, tailor-made treatment after meeting a Knome doctor who flew out to explain what he might learn during the process. After wiring the 50 percent down payment, a clinical specialist drew his blood.[8] His DNA was isolated in a New Jersey lab, but the actual DNA sequencing was outsourced to the little-known Beijing Genomics Institute (BGI) in Shenzhen, China. After Conde read press reports of the first Asian genome sequence, he visited BGI and was blown away by the raw technology and manpower on display, not to mention the competitive pricing. The Chinese were using Illumina machines, but Conde didn't care what platform they used as long as the sequence quality and pricing were acceptable.

The final step was to invite the client to Boston to receive a guided tour though his genome. During a day-long roundtable session, Knome's geneticists and clinicians presented the highlights of his sequence analysis. While Knome had outsourced the DNA sequencing, its key strength was its homegrown genome analysis software, a powerful yet easy-to-use interface providing clients with an unsurpassed interpretation of their DNA. Conde called it the Rosetta stone of personal genome analysis. Together, the team looked at the common SNPs, rare or private mutations, and Mendelian disorders, paying particular attention to actionable conditions. The dramatic highlight of the presentation came when Conde, like a banker in a James Bond movie, unlocked an attaché case and presented the client with a titanium USB drive containing his complete digital genome. Conde explained: "Their assembled sequence and the genome are loaded onto this genome key. It talks to us. They can review their data offline. They don't have to log in anywhere." If the client wanted an update, Knome pushes any new results onto the key. "The client is not giving their genome to us to store for them, so they maintain control over their information." And if the client erroneously entered the password ten times? The drive self-destructs, just like an old episode of *Mission Impossible*.

Following Stoicescu as one of Knome's early clients was University of Illinois professor and novelist Richard Powers, who was encouraged by the editor of GQ magazine to write about his genome. Powers hesitated—he was in the middle of writing his novel *Generosity*—but eventually agreed. Knome and GQ agreed to a special cut-price draft version of Powers's genome.[9] Powers still had doubts as he digested the Knome consent form, which warned of risks to employment or insurance status should his genetic or medical information become public, and reiterated that Knome's services were intended "for informational and research purposes only." And it continued:

> NEITHER KNOME, NOR ITS DIVISIONS, SUBSIDIARIES, SUCCESSORS, PARENT COMPANIES, AFFILIATES OR THEIR EMPLOYEES, PARTNERS, PRINCIPALS, AGENTS AND REPRESENTATIVES, NOR ANY OTHER PARTY INVOLVED IN CREATING, PRODUCING OR DELIVERING WHOLE-GENOME SEQUENCING, ANALYSIS, INTERPRETATION, OR COUNSELING IS LIABLE FOR ANY DIRECT, INCIDENTAL, CONSEQUENTIAL, INDIRECT, PUNITIVE OR ANY OTHER DAMAGES ARISING OUT OF SUCH SERVICES OR YOUR USE OF THESE SERVICES. THIS INCLUDES LIABILITY FOR PERSONAL INJURY OR DEATH.

Powers received his titanium flash drive in the auspicious surroundings of the Harvard Club in downtown Boston. Joining Conde were Church and his Harvard Medical School colleague Raju Kucherlapati, and Hugh Rienhoff, physician and founder of MyDaughtersDNA.org, Powers was told he had 1.3 million variants, insertions, and deletions in his DNA compared to the reference genome, including more than 51,000 novel variants that scientists had never documented before. "Each new whole genome is something of a virgin continent," Powers wrote. Six hundred of those novel variants were tagged as potentially deleterious, of which 40 percent had a putative disease associated with them. Conde informed Powers he had the so-called novelty-seeking gene—described in a pair of high-profile studies I published in *Nature Genetics* in 1996 linking a particular version of the dopamine receptor *DRD4* gene in thrill seekers, bungee jumpers, and the like.[10] The media lazily reported this discovery as *the* novelty-seeking gene, even though the small print acknowledged there were probably one hundred or more contributory genes. Powers also learned he carried risk factors for obesity, although his family nickname, "stick man," suggested this was not a major concern. While Powers didn't learn a great deal of immediate medical relevance, another early Knome client found he had a couple of rare mutations associated with glaucoma. Although he was asymptomatic, his ophthalmologist uncovered early signs of intraretinal pressure probably a few years earlier than would have otherwise been the case.

Early on, Conde predicted that Knome's price would fall steadily and approach the $1,000 range within ten years. But he had underestimated the pace of change. At $150,000, a Knome personal genome was still a luxury item, but it was a whole lot cheaper than a Bentley. By the end of 2008, Knome had recruited twenty clients from around the world, aided by the sinking dollar. After a further price drop to $99,000, Knome signed up five new customers in the first week, largely through word of mouth. "We're a small company," said Conde. "We're not going to buy an ad at the Superbowl or anything like that." Then he reconsidered. "Well, maybe next year!" Following the 2009 DNA Day eBay auction, Knome set its price at $68,000, but with the ever improving technology from Illumina and Applied Biosystems, it wouldn't stay there for long.

By early 2010, reports of more than a dozen human genomes had been published in leading science journals, and dozens more had been completed. Those of Venter and Watson were joined by two cancer patients; a Yoruban African (twice) and a Han Chinese; two Koreans; Stephen Quake (on his own Helicos instrument); and three by Complete Genomics, including George Church. There soon followed Baylor geneticist Jim Lupski, a member of the Project Jim team; Archbishop Desmond Tutu and four African Kalahari Bushmen; and a pair of Utah siblings and a Turkish child with rare genetic disorders. Several others had been announced ahead of publication, including a Dutch clinical geneticist, the first Arab genome, the first Indian subject, a Russian, and more cancer patients. Complete Genomics delivered its first fourteen genomes to clients, with orders for 500 more. Many more had been done as part of the 1000 Genomes Project, a cancer genome project, and the Personal Genome Project. Illumina had launched its own personal-genome sequencing service, similar to Knome's, with CEO Jay Flatley, venture capitalist Hermann Hauser, and Harvard professor Henry Louis Gates and his father getting the ball rolling. Orders arrived quickly, including former Solexa CEO John West's family of four. Not to be outdone, Life Technologies CEO Greg Lucier joined the whole-genome club as well.[11]

Given the sudden rush and diversity of human genomes being reported, it was probably only a matter of time before Hollywood got involved. In March 2010, Illumina announced it had sequenced the genome of actress Glenn Close, a committed activist on behalf of mental illness.[12] While Close became the first Oscar winner to be sequenced, she was not, as some commentators assumed, the first named woman to do so. Shortly before the Watson press conference in May 2007, Gert-Jan van Ommen and some colleagues from the University of Leiden Medical Center in the Netherlands were discussing the prospects of human genome sequencing. Van Ommen had taken delivery of the first Solexa machine in continental Europe and, after practicing on pond snail DNA, wanted to try something more ambitious—an individual human genome. "That was reasonably naive" back then, he admits, but his group did hold the record at the time for Solexa sequencing throughout Europe.[13] Stefan White, a PhD student from New Zealand, said, "They already did Watson, so why don't we do Kriek?" He meant Marjolein Kriek, a red-headed, freckle-faced clinical geneticist in the group. Her surname, pronounced in English, would be "Crick"—as in Francis Crick, Watson's double helix partner.

The next day, Johan den Dunnen, head of the Leiden Genome Technology Center, invited Kriek to become not only the first European to be sequenced but also the first female. (One reason scientists tend to prefer male genomes is that women don't carry a Y chromosome.[14]) Kriek reflected on her family obligations, not least the fact that she was pregnant, but ultimately agreed—with one caveat: "I'm going to block the mismatch repair genes for colon cancer, not because I don't want to know, but there might be insurance complications," she told me.

With no funding for the project, the Dutch team simply stuck in some of Kriek's DNA whenever there was some spare capacity on the Solexa machine. By spring 2008, they had compiled some 22 billion bases for an average sevenfold coverage, although the quality was poor.[15] Still, van Ommen decided to use the annual Bessensap meeting of Dutch scientists and journalists in May 2008, held at the Science Museum in Amsterdam, to announce Kriek's genome. "We could see that Marjolein is a female, because there is no Y chromosome DNA," van Ommen joked, but little other detail was presented. The cost was estimated at about 40,000 euros.

From Amsterdam, van Ommen traveled to Barcelona for the annual gathering of Europe's human geneticists. "Illumina was jumping with joy," he recalled. So too was the creator of the popular Dutch newspaper comic strip *Fokke & Sukke,* the musings of a politically incorrect duck and canary, respectively. "Marjolein Kriek??!" says Fokke incredulously in one comic strip, to which Sukke replies: "Why not unravel the complete DNA of Scarlett Johansson?!"[16] Kriek was also honored in a museum exhibit to mark Darwin's 150th anniversary: a life-size photograph of Kriek with her DNA sequence projected onto her dress. Even more exciting, she exclaimed, "I've got my own statue!"[17] In a square in Rotterdam, alongside a statue of Erasmus, is a four-meter-tall statue containing portions of her sequence printed on a stack of sailing cloths. Now, when van Ommen lectures to medical students and asks about the genome pioneers, they usually remember the pretty redhead rather than Watson and Venter.

Since the 2008 announcement, the Leiden team has continued sequencing and assembling Kriek's genome, but "the venom is in the tail," said van Ommen. "What's caught people's attention much more than the As, Cs, Ts and Gs is, why does somebody do this? Aren't you scared that you'll find things out?" Kriek takes that responsibility seriously, as she envisions people arriving at her clinic and asking her to interpret their genome sequence. She wants to make the public less afraid of what they might find. "They don't know me when they know my sequence," she said. She is now the doting mother of a baby girl, Guna, but has no interest in sequencing her daughter. "She has a perfect genome!" says Kriek sweetly. "I think a person should be old enough to make her own decision. . . . Because she might say, 'Mama, nice you did that, but don't make decisions for me.' It'll be there for her whole life."

While 454 Life Sciences had the distinction of producing the first next-gen genome, Illumina rounded off a banner 2008 by hosting a press conference in Philadelphia to mark the publication of three landmark articles in *Nature.*[18] Illumina's Solexa sequencing was the workhorse behind all three papers: on the first African genome sequence, the first Asian genome, and the first cancer genome.

The first speaker was the leader of the Chinese group that had sequenced the first Asian genome, a Han Chinese. (Han is the largest ethnic group in China, making up about 90 percent of the population.) Only in his mid-thirties, BGI Shenzhen's Jun Wang was a Chinese scientific rock star, though he would cringe at being described as such. Tall, handsome, speaking in fluent English, Wang said "People are doing this because they are trying to predict their life." Even in China, Wang said, there were already unreasonably high expectations for human genome sequencing. "I'm often asked, how long will the guy live?" he said. Wang did not identify the subject: his YH prompted speculation that he was BGI director Yang Huanming, but such a disclosure would be considered unethical and this has not been officially confirmed.[19] Nevertheless, "This guy is a pure Asian," confirmed Wang. "He has a very severe tobacco addiction. This matches the genotype information very well." Wang explained the reason for sequencing an individual's complete genome: "Before, we studied candidate genes—like trying to find keys in a dark street. Now we're lighting up the whole street."

Wang had been a key player in BGI's efforts to contribute 1 percent of the Human Genome Project and 10 percent of the international HapMap project a few years later. But the Chinese wanted their own Asian reference genome, not just for national pride but to support future medical genetics studies, and now they had it. Wang's team first announced the sequence at a Chinese technology fair in September 2007—just the third personal genome after Watson and Venter. The BGI team found 3 million SNPs, including 400,000 novel variants, 135,000 short insertions or deletions, and 2,600 larger structural variants. There was so much data that the analysis took twice as long as the sequencing. "Today we could get the sequence done in about one week," Wang said, and for just a few thousand dollars. Obviously this would not be the last. "One human genome is *not* enough!" he said. "We're going to sequence many, many more individuals," as he flicked to a slide of a Shenzhen beach packed with sunbathers.[20]

Among the samples being sequenced at BGI was a genome from an unlikely source. In 2007, Prince Ahmad bin Sultan bin Abdulaziz, the son of the crown prince of Saudi Arabia, established Saudi Biosciences to serve as scientific counselors to the kingdom. Saeed Hussain of Saudi Biosciences noted that, "Arab populations were not participants in the HapMap or the British Genome Project."[21] Prince Ahmad believed an Arab genetic map was a crucial first step in identifying the roots of complex genetic diseases in the region, beginning with diabetes, which affects one in four Arabs due to various environmental and genetic factors, or what Hussain quaintly calls "the cousin marriages." The project calls for the selection of one hundred Arab volunteers, according to Hasidic and genetic studies, from five regions with relatively homogeneous populations, from Jordan and Syria to Yemen, Egypt, and North Africa.[22] "Arabs have occupied most of the migration routes from Africa to India, through Egypt to central Asia, Europe. It would be lovely to understand the genetic makeup from Africa to Arabia to central Asia to Europe," said Hussain.

The Saudis soon learned what everyone else knew: sequencing

was the easy part. In September 2008, Saudi Biosciences announced it had finished the first Arab genome sequence—a feat performed in China with British chemistry and using software from Denmark. Hussain said proudly: "This project launches the Kingdom of Saudi Arabia into the small circle of nations who are currently in the process of building sophisticated databases of human genetic variation." The goal was to complete the first one hundred Arab genomes by the end of 2010, but Hussain said the ultimate objective was developing prognostic kits for genetic diseases.

Wang's ambitions did not stop with Western millionaires, Arabian royalty, or Shenzhen sunbathers. "If it tastes good, we'll sequence it," he said, showing pictures of a cucumber, rice, a tomato, and a pig. "If it looks cute, we sequence it"—and he showed a panda, polar bear, and emperor penguins. "We want to know why it looks so cute!" In 2009, Wang's team sequenced one of the cutest: the genome of a Jingjing, a three-year-old female panda who was a mascot of the Beijing Olympics. The panda has 2.4 billion bases in twenty-one pairs of chromosomes, but, curiously, no recognizable genes for the enzymes that break down cellulose, suggesting that bamboo digestion may be aided by bacteria in the gut rather than the panda bear biochemistry.[23]

To make good on his ambitions, Wang would need some serious artillery. In early 2010, reinforcements arrived: with a massive $1.5 billion loan from a major Chinese bank, BGI ordered 128 (considered a lucky number) of the latest Illumina HiSeq machines, each capable of sequencing two human genomes side-by-side in a week. By the end of 2010, Illumina CEO Jay Flatley predicted that BGI's new sequencing center in Hong Kong would have a sequencing output greater than all of the genome centers in the United States.[24]

The second Illumina success story at the close of 2008 was described by Rick Wilson, director of The Genome Center at Washington University in St. Louis. The first cancer genome belonged to a white female in her mid-fifties, who had acute myelogenous leukemia (AML) and died before the report came out. Interestingly, her tumor DNA showed no signs of the rampant chromosomal rearrangement or loss of copy number so common in cancers, and although a handful of suspicious mutations had been seen, it was too soon to tell if they were cause or effect. A second AML genome also sequenced by the St. Louis team that later appeared in the *New England Journal of Medicine,* belonging to a male in remission after receiving chemotherapy, provided a breakthrough.[25] This genome took just 16.5 runs on Illumina machines to obtain near-complete (98 percent) coverage of the cancer genome.[26] "At current rates, we really can sequence a genome a week," said team member Dan Koboldt.[27] The most striking result was a recurring mutation in a gene called *IDHI,* observed in 16 out of 187 AML samples, which offers a range of potential diagnostic and therapeutic options.

Closing the briefing was Scott Kahn, Illumina's chief information officer. Illumina's own *Nature* paper on the first African genome put the list price of the reagents at $250,000, somewhat higher than the $100,000 claimed in an earlier press release. Kahn said the three papers made a powerful statement about how far the technology has come: "I don't think anyone in this room would have thought that, in a single publication in *Nature* in 2008, there'd be effectively four genomes published. That is just mind boggling."[28]

The next two genomes published were Korean. The first in print belonged to a team at the Lee Gil Ya Cancer and Diabetes Institute.[29] The original plan was to sequence university president Lee Gil Ya as a golden anniversary gift. But when the pilot data were presented at a medical conference, Gil Ya faced "enormous objection from her friends and staff," worried about the potential disclosure of health problems, said project leader Sung-Min Ahn.[30] Instead, the Koreans sequenced Seong-Jin Kim, the director of the institute, using the first Illumina instrument in the country.[31] "Koreans and the Chinese are very close, yet the two individual genomes from them showed quite a big variation," said Ahn. Indeed, almost 6 percent of Kim's DNA could not be mapped to the reference genome, underscoring the surprising range of genetic diversity from different populations.

The second Korean paper was, in fact, the official Korean Genome Project, from the Seoul National University College of Medicine.[32] The DNA donor, AK1, was a thirty-year-old "healthy and intelligent" male who had donated blood eight years earlier for Korea's early genomics effort. The sequencing took six weeks, the analysis twelve. Using George Church's Trait-o-matic program, they learned AK1 might develop "reduced sensitivity to statins, and tuberculosis susceptibility in the future," said study director Jeong-Sun Seo. The Koreans claimed to have produced the most detailed analysis of structural variants for any genome thus far and, at an average of 106 bases, the longest read length for any next-gen genome (except Watson). The next step was the Asian 100 genomes project, a Noah's ark approach of selecting at least two individuals from each country in northern and southern Asia. The work extended a project to studying isolated tribes in a remote region of Mongolia, cleverly named GENDISCAN (Gene Discovery for Complex Traits in Asians of Northeast), with the goal of establishing an Asian-specific genome database for personalizing medicine in Asian populations.

In the summer of 2009, Kevin McKernan's team at Life Technologies (the company formed by the merger of Applied Biosystems and Invitrogen) published data on the first human genome sequenced on the SOLiD platform. McKernan, Gina Costa, Alain Blanchard, and colleagues had sequenced NA 18508, a Yoruban genome, in 2007 and 2008. It was the same HapMap sample that Illumina had published in 2008, but while McKernan's paper cited the earlier study, there was no direct data comparison. One reason, he said, was that Illumina genotypes (the identities of some 4 million SNPs) were not available when he submitted his paper.[33] With everything else they were doing, including developing SOLiD 4, McKernan said it was "hard to get folks jazzed about a retrospective bake-off," particularly when his ABI team was generating up to 50 billion bases per run and aggressively shooting for 300 billion by the end of 2010. Amid some doubts about SOLiD's prospects against Illumina, McKernan shot back: "I don't think this race is over in the first 900 machines, considering over 12,000 [ABI Sanger] instruments have been sold."[34]

Amid the first wave of personal genomes, no other story captured the medical imperative and potential of genome sequencing as that of Hugh Rienhoff and his daughter, Beatrice. When he quit DNA Sciences in 2001, Hugh Rienhoff probably thought his involvement in personal genomics was done. "My daughter pulled me back into all of this," he told me, in something of an understatement. From the moment he first held his baby daughter, Beatrice, in 2003, Rienhoff sensed something was amiss. Noting her unusually long feet and clawed fingers, Rienhoff, who had trained at Johns Hopkins with the father of modern medical genetics, Victor McKusick, worried his daughter might have inherited Marfan syndrome, which McKusick had suggested affected Abraham Lincoln.[35] McKusick had created the essential reference catalog of genetic traits and disorders called Mendelian Inheritance in Man (MIM), which Rienhoff likened to the first edition of the *Oxford English Dictionary*.[36]

Beatrice failed to gain weight and was later hospitalized for failure to thrive. Doctors suggested a host of medical possibilities and diagnoses, but nothing panned out.[37] Almost as a last resort, Rienhoff flew his then eighteen-month-old daughter to Baltimore to visit David Valle, director of the Institute of Genetic Medicine at Johns Hopkins, who set up a makeshift clinic. Rienhoff watched the residents nod with quiet conviction as they noticed something that had been inexplicably missed: Beatrice had a forked uvula (the flap of skin in the back of the throat). Just weeks earlier, two of Valle's colleagues had published a report describing the genetic basis of a disorder named Loeys-Dietz syndrome.[38] Patients had a split uvula and, much more serious, a potentially fatal weakness of the aorta, similar to Marfan syndrome. Bart Loeys gave Rienhoff a copy of his *Nature Genetics* paper to read on the plane back to California.

Rienhoff read the article with a sinking feeling. The average life span of Loeys-Dietz patients was just twenty-seven years. He had probably seen many cases of Loeys-Dietz during his Hopkins residency and misdiagnosed them as Marfan's. In the event, the sequencing tests on Beatrice's DNA for the errant gene, the TGF-ß receptor, came back negative, but Rienhoff had the bit between his teeth. Turning his attic into a makeshift genetics studio, Rienhoff doodled cellular signaling pathways that might explain Beatrice's mysterious muscle weakness, focusing on myostatin, a hormone closely related to TGF-ß that regulates muscle fiber growth.[39] Could Beatrice carry a hidden flaw in one of her myostatin receptors that stunted her own muscle development?

Rienhoff spent an afternoon in 2006 in the lab of Stanford professor Andy Fire, who had just won the Nobel Prize, extracting his daughter's DNA from her blood. He then invested $2,000 in a used polymerase chain reaction machine and other lab equipment. "If you can make a good soufflé, you can sequence DNA," he joked,[40] although in the end, he elected to send off his daughter's DNA to a company for some targeted sequencing of a candidate gene. With the results in hand, he painstakingly searched the 20,000 bases for any glitch that could account for his daughter's disorder. In one spot, Beatrice had a G instead of an A, a change never reported at that position before. But Rienhoff's hopes were dashed: it turned out he carried the same benign variant. Still, the DNA detective work did not go to waste. He persuaded his daughter's cardiologist to prescribe a blood pressure drug, losartan, that showed positive effects in animal models of Marfan syndrome. Rienhoff's greatest worry was that any weakness in Beatrice's major blood vessels could get worse if unchecked, whereas a therapy could always be halted if any adverse symptoms arose.

Rienhoff decided to publicize his daughter's struggle by launching a Web site to inspire others searching for clues to undiagnosed conditions in their loved ones—MyDaughtersDNA.org. Harvard neuroscientist David Clapham, for example, movingly described the death of his nine-year-old son on his lap from an undiagnosed neurological condition as "like winning the lottery in reverse."[41] "There's no place where people and geneticists share unsolved cases," said Rienhoff. "It might be quixotic to think that physicians might take time to post cases. . . . Physicians don't really understand genetics and they're too busy to learn." But they have, and in at least one case, a successful diagnosis has been made.[42]

In October 2007, Rienhoff and his daughter graced the cover of *Nature*, a tribute to his determination and her considerable courage. "With a sequencer and a website, Rienhoff has stepped over the threshold of personal genomics in a way set to catch the imagination," wrote Brendan Maher. "As sequencing gets ever easier and knowledge bases ever larger, it may not be fanciful to imagine more and more people following him, developing theories about abnormalities and testing them through sequencing. Such attempts will often fail, and in some cases lead to frustration and heartache. But some may make significant contributions to our understanding of the function of various human genes."[43] With help from George Church, Rienhoff is also laying out his daughter's medical history in a phenotype spreadsheet. Rienhoff believes such representations will one day enable a computer to systematically analyze a patient's symptoms with the encyclopedia of genetic disorders.[44]

Rienhoff's self-help approach was not universally appreciated by his peers, however. At one symposium, he was chided for his hands-on methods. "Who the —— is this person?" Rienhoff recalled thinking to himself, dismayed by the contempt a doctor had for the curiosity of her patients.[45] And Johns Hopkins's Hal Dietz advised Rienhoff to remove his do-it-yourself instructions for DNA sequencing from his Web site. Rienhoff acceded, but he still sees a revolution brewing. "Doctors are not going to drive genetics into clinical practice. It's going to be consumers," he said. "The user interface, whether software or whatever, will be embraced first by consumers, so it has to be pitched at that level, and that's about the level doctors are at. Cardiologists do not know dog shit about genetics."

At times Rienhoff blames himself for literally spawning the mutation that he passed onto Beatrice at conception. As he said during a talk at Google, "Dads—particularly that are forty-five and over—they throw off a lot of bum seed."[46] What he said then applies still. "My primary charge as a physician, and as a dad, is to do what I can." And if that meant sequencing his daughter's entire genome to understand once and for all the molecular basis for her mystery condition, then that's just what he would do.

In August 2009, Stanford professor and Helicos cofounder Stephen Quake published a paper on the sequencing of his own genome,[47] which he had first revealed in a *New York Times* op-ed earlier that year.[48] Working with just two colleagues, Quake said this was "the first case in which you haven't needed a genome center to sequence a human genome." Moreover, it was the first report of decoding a human genome by single-molecule sequencing and a timely shot in the arm for embattled Helicos. One person who felt vindicated was Kevin Ulmer, who befriended Quake while working in Steve Chu's lab at Stanford in 1993 and founded the first single-molecule sequencing biotech, SEQ.[49]

Because of conflict-of-interest rules,[50] Quake couldn't simply buy the $1 million instrument from his own company, but when his colleagues at the Stanford stem cell institute acquired a HeliScope for its own studies, Quake jumped the queue to volunteer his DNA. It took research associate Norma Neff just four runs on the HeliScope, each lasting a week, generating 148 billion bases at an estimated cost in reagents of $48,000. Quake's PhD student, Dmitry Pushkarev, handled the computational analysis. The reads were short, less than half the length of Illumina, and a high error rate (about 3.5 percent) meant that fewer than two-thirds of the reads could be aligned to the reference. More than half of the errors were deletions resulting from the "dark base" issue that continued to frustrate Helicos.[51]

Using Church's Trait-o-matic software, Quake found a rare mutation associated with a heart disorder, for which there may be some family history. "If you know your uncle had something, you kind of discount that you can get it, but to see you've inherited the mutation for that is another matter altogether," he said. He also carries a variant in the *CLOCK* circadian rhythm gene, which might be associated with increased disagreeability. "You don't need my genome to tell you that," Quake quipped. A much more detailed appraisal of the clinical ramifications of Quake's genome was published nine months later. A small army of Stanford physicians exhaustively pored over Quake's sequence, taking particular note of his family history of cardiovascular disease, including the loss of a teenage cousin due to a suspected sudden heart attack. At least three genes contained rare and potentially "damaging" mutations that might increase Quake's chances of developing coronary artery disease and diabetes.[52] Quake promptly sought a cardiac ultrasound, but he was still unsure whether he wanted to start taking prescribed statins and be a patient the rest of his life.

Quake felt his success marked a tipping point in genome research—not to mention the fortunes of Helicos, which he portrayed as David in a battle against three $1 billion Goliaths. His friends at Helicos were "a scrappy little bunch, trying to hang on. I think they're fantastic, and I'm hoping they're going to end up at the top of the heap." What particularly pleased Helicos chairman Stan Lapidus was that "compared to the usual army of people involved in sequencing a genome, the fact that so few were involved here is itself a breakthrough and foreshadows the future of sequencing." The day was fast approaching when sequencing complete genomes would be as routine a lab procedure as loading the sample and pressing a button.

The Quake study drew some mixed reviews, however. One genome center official dismissively chalked the work up as just "another rich white guy's genome."[53] Another blogger said the quality of the platform was no better than the studies of Illumina twelve months earlier. What particularly irked observers was the dubious claim that somehow the HeliScope was not only cheaper but easier to operate than the competition. "They're using the number of names on the Solexa [2008 *Nature*] paper as evidence of how many people are required to run an instrument," complained an incredulous Clive Brown, Solexa's former software chief.[54] "That Solexa paper was the culmination of eight years work. It had everybody's name on it. The CEO's name was on there, and he didn't even do anything!" He added that the original Helicos paper required twenty-three authors just to sequence a virus. But overall, he thought Helicos deserved some kudos for sticking to its guns.

Quake wasn't the only scientist with connections to be curious about his own genome. On June 10, 2009, Jay Flatley announced the launch of Illumina's personal-genome sequencing service, priced at $48,000. The Illumina CEO said he had dreamed of pushing into personal genomics since he bought Solexa in 2006.[55] "It was just a matter of what was the right time in terms of the market and our technology." Illumina's Personal Genome Sequencing Service would be performed in the firm's CLIA-certified laboratory, but only with a doctor's note. Technically it wasn't quite "direct to consumer," but Flatley was treading cautiously, not interested in courting controversy. After potential clients are briefed by their doctor and signed the consent form, they'd go into a "cooling-off period" for seven days to ensure they want to proceed. Once the blood sample is delivered to Illumina, the customer pays 50 percent of the fee. Illumina sequences the genome to thirtyfold coverage and then delivers the sequence to the physician, who communicates the results to the consumer. (The genome was delivered to the consumer on an iMac computer running Illumina's GenomeStudio software.)

The genome data would include information on SNPs, structural variations, and insertions and deletions, but not any medical interpretation. That would be provided by any of the consumer genomics firms, including Knome. "It's not Illumina's intent, nor is it our skill, to connect genetic information to medically relevant information. That's a role we're going to ask other companies to help us play," said Flatley at the launch.

Flatley proudly displayed a photograph of the world's first prescription for a human genome sequence, issued in January 2009 by University of California San Diego medical geneticist Robert Naviaux after Flatley's physician politely declined. The other three volunteers for the $48,000 special were Hermann Hauser, a venture capitalist with Amadeus Capital Partners who had been an early backer of Solexa;[56] Harvard professor and film producer Henry Louis Gates Jr., who was working on the PBS series *Faces of America*; and Gates's father. Flatley wasn't expecting a stampede at that price, but the falling costs made it feasible to dip a toe in the water: "We think it's time for this process to begin. By providing this service now, Illumina can help catalyze the development of the infrastructure and physician education that will be necessary as genomic information becomes medically more meaningful."[57]

Not content with being an early adopter of genome sequencing, Flatley had one of his software developers whip up a personal-genome app for his iPhone and later his new iPad. The app would search data on various diseases and traits, drug response, and chromosomes or genes and share information with friends and family. He declared his own genome reasonably boring, save for a large deletion on chromosome 22. "My wife believes it's the empathy gene," he cracked. In June 2010, on the first anniversary of its individual-genome sequencing service, Flatley announced a dramatic price cut to just under $9,500 for a clinical case and $19,500 for an individual.[58] The only catch is that Illumina no longer throws in the free iMac. Preliminary analysis of Flatley's own genome revealed 64 potentially "disease-causing" mutations, based on the annotation in the public databases. Fortunately for Flatley not all are accurate: the entry for one of these variants stated, "Death in early infancy highly likely." Upon further review, the majority of the suspected mutations were invalidated, but Flatley said he carries 11 confirmed variants associated with various genetic disorders, most of which he had never heard of. It was clear evidence, if proof were needed, how all of us carry a significant number of genetic glitches in our genomes.

• • •

The year 2009 ended with clear signs that whole-genome sequencing was not just an amazing research tool but capable of delivering medical benefit as well. Baylor's Jim Lupski is a pioneer in the study of structural variants underlying genetic disorders. In 1991, his team documented an unusual duplication of part of chromosome 17 associated with a nerve disorder called Charcot-Marie-Tooth (CMT).[59] (Such structural rearrangements affecting the copy number of groups of genes didn't get full recognition until their rediscovery in 2004.) In two decades, Lupski and others cataloged CMT mutations in some forty genes, but none appeared to correlate with the disorder in one particularly interested patient: Lupski himself, who has undergone a string of orthopedic operations. In an effort to solve the mystery once and for all, Baylor's Richard Gibbs offered to sequence Lupski's entire genome. Using the SOLiD platform, Gibbs sequenced Lupski's DNA, then whittled down the millions of SNPs in search of putative disease-causing mutations.[60] Much to Lupski's surprise and relief, the gamble paid off. "I've been trying to figure this out for a long time," Lupski told me.[61]

When Lupski focused on just the known CMT genes and others linked with various neuropathies, there was a short list of fifty-four candidates. One of those SNPs tracked in Lupski's father and grandmother, both of whom had some CMT-like symptoms. That gene was *SH3TC2*, which encodes a protein likely to have a role in the insulation of nerve fibers. Further sequence analysis revealed a different mutation in Lupski's other copy of the same gene. Lupski's intimate knowledge of the biology of CMT undoubtedly facilitated the identification of the gene culprit. "It was only because of that clinical knowledge that we could make this tie," he said, or he might still be looking. Some argued it was ridiculous to sequence Lupski's entire DNA when a more systematic search of all the candidate genes would have netted the same result. But several other studies have reinforced the power of sleuthing disease genes by whole-genome sequencing.

Two research teams identified the genetic basis of an extremely rare hereditary disorder in a small family, the Madsens of Salt Lake City, which thus became the first named family to have their genomes sequenced. Debbie Jorde's two children, Logan and Heather Madsen, were two of only some thirty patients worldwide known with Miller syndrome, characterized by facial and limb abnormalities. Not only did two teams of researchers identify the Miller syndrome mutation (Debbie's second husband, Lynn, a noted author of human genetics textbooks, coauthored one of the studies), but even more remarkable, they also uncovered the mutation behind a second genetic disorder affecting the Madsen siblings—a cystic fibrosis–like disease called primary ciliary dyskinesis. (The odds of having both were put at less than 1 in 10 billion.) One study was noteworthy for featuring Complete Genomics, an important validation for the new sequencing platform. Moreover, by comparing the family's genome sequences, one of the teams produced a new calculation that each parent had passed down about thirty mutations to their children, less than the conventional wisdom. But Debbie Jorde was focusing on a higher purpose. "The possibility that we can be of help to other people, humankind, really makes us feel like living with all of these challenges hasn't been a waste, that it was for a reason," she said.[62] And Rick Lifton's team at Yale University uncovered by exome sequencing the mutated gene in a Turkish infant suffering from a mystery life-threatening kidney disease.[63]

After devoting two decades to the study of genetic disorders such as CMT, Lupski could enjoy a brief moment of satisfaction. But as he accepted the congratulations of friends and colleagues after his talk at the American Society of Human Genetics convention, I could see he was still sore at *Nature* for returning his paper unreviewed. Apparently sequencing a complete human genome and finding disease-causing mutations wasn't what it used to be.[64]

Not that *Nature* was done publishing tour-de-force sequencing studies. At the same meeting, Penn State's Stephan Schuster announced the sequencing of a Kalahari Bushman, representing possibly the oldest human ethnic group in the world. The Bushmen are nomadic hunter-gatherers, small in stature, who do not practice agriculture or herbiculture. Schuster and his colleague Vanessa Hayes gained the trust of these largely forgotten Africans, about 70,000 to 100,000 and living mostly in Namibia and Botswana. While indigenous populations have cause to be suspicious of researchers' motives, the Bushmen agreed to collaborate, not only to demonstrate their unique place in human history but also in the hope that new knowledge would provide future medical benefits.

Schuster received full backing from Roche-454, the platform he'd used for his ancient DNA studies, but a nonfactor in complete human genome sequencing since Project Jim. He found that the Bushmen fell into the L0k and L0d mitochondrial (mt) haplotypes, among the deepest branching groups, about 150,000 years old. While there are some twenty mtDNA differences between the Bushmen and Europeans, the diversity among the Bushmen is around seventy to eighty nucleotides between individuals—a clear indicator of the age of the population. That conclusion was buttressed by analysis of the full genome, which revealed that different Bushmen linguistic groups harbor more genetic differences between each other than a European and an Asian. "That is incredible," said Hayes.

In February 2010, Schuster and Hayes hosted a press conference in Namibia, joined by the Namibian prime minister and a special guest of honor, Archbishop Desmond Tutu, whom Hayes and Schuster had also sequenced. Tutu is of mixed Bantu ancestry, the dominant ethnic group in southern Africa, but was delighted to learn that he also carries some Bushman genes. The Nobel Peace Prize winner is also a survivor of TB, polio, and prostate cancer. Hayes expressed hope that the study would dispel the traditional Eurocentric view of genetics research, while Schuster hoped to spotlight the need to assist indigenous groups in their struggle against loss of land, famine, and disease.[65]

Meanwhile, Rade Drmanac and colleagues at Complete Genomics were making rapid progress in building up their genome sequencing service and published a progress report in *Science*. Twelve months earlier, *Nature* had published three back-to-back articles describing three human genomes. On New Year's Day 2010,

Drmanac described the sequencing of three human genomes in a single paper, including that of PGP-1, George Church. "George's genome is the cheapest sequenced human genome on earth," he told me,[66] although the estimate didn't include the cost of space or electricity, manpower, or instrument depreciation. The average reagent cost per genome was $4,000, but Church's genome was sequenced for just $1,500—another indisputable signal as to how routine and commonplace human genome sequencing had become in the space of just a couple of years. In fact, the Complete article barely scraped into the top five stories listed in that week's *Science* press release. Complete Genomics CEO Cliff Reid told me that his company was routinely sequencing genomes for reagent costs under $1,000.[67]

In an appearance on the *Charlie Rose Show*, George Church seemed surprisingly coy when the host asked him about his own genome. "I actually haven't studied my own [genome] nearly as much as the other nine [PGP volunteers]," he said.[68] "Nothing floated up to the top of my list that was worth spending much time on." I asked Church whether, like Eric Lander and Kari Stefansson, he was more comfortable perusing other people's genomes than introspectively searching his own. "I think I land in between Eric Lander and Craig Venter," he said. "Craig is clearly not shy about his genome—he wrote a whole book [about it]. I'm doing full disclosure—maybe more than anybody else—I'm not obsessed with it."[69]

One of the unsolved mysteries in Church's sequence is the genetic basis of his narcolepsy, a chronic sleep disorder. Although a mutation for narcolepsy has been identified in some breeds of dogs, only one human case with the corresponding glitch has been documented. Most cases in humans appear to be the result of an autoimmune disorder, reducing the number of neurons in the brain releasing a "wakefulness" peptide. It turns out Church doesn't have this gene variant, but his wife and daughter, who shows narcoleptic symptoms as well, do.[70] "It could easily be that my daughter got a double whammy," he said.

For Church, making sense of that data would take months, if not years, and go far beyond simply the stamp collecting exercise of cataloging all the gene variants and mutations. Church regarded the genome sequence as just the beginning: he envisions that for each genome, scientists will build computational models, analyze patterns of gene activity, and start reverse-engineering specific gene variants into cells to look for causation rather than merely correlation. That all-encompassing set of data, not just the DNA we inherit but a reading of all the extrinsic factors that could influence or harm its sequence and function, would enable the personal prioritization of clinical diagnostics, therapeutics, lifestyle, and nutritional changes.

The tenth anniversary of the first public declaration of the cracking of the human genome was marked in June 2010. What once took a decade and billions of dollars can now be produced routinely in just over a week for a commercial price of less than $10,000, and a research cost of $1,000. It has already reached the point of a routine event that is rapidly moving from the early adopters and pioneers into the hands of anybody who wants it.[71] When a Russian group published details of the genome of the first ethnic Russian, using both SOLiD and Illumina systems, it barely registered a raised eyebrow.[72] It's hard to know what is the more remarkable: how mainstream affordable human genome sequencing has become or how much faster and cheaper the technology can go.

CHAPTER 11

The 15-Minute Genome

In 2010, Illumina and Life Technologies were vying for dominance in second-generation sequencing. With its new HiSeq 2000 machine, Illumina promised researchers the ability to sequence two genomes side-by-side in eight days for about $10,000 a genome. Kevin McKernan's team at Life Technologies was shooting for 300 billion bases of sequence in a run, driving the cost of whole genome down from $6,000 to just $3,000 by the end of the year. Meanwhile, a host of groups, from giant corporations to bootstrapped start-ups and idealistic academics, were looking ahead to the true killer app: the third-generation sequencing system that would not only deliver the Holy Grail, the vaunted $1,000 genome, but push beyond to the $100 genome or even cheaper.[1]

Some schemes, such as the one developed by former nuclear power engineer William Glover at ZS Genetics in New Hampshire, opted to simply stretch out DNA molecules studded with heavy atoms such as iodine and bromine that render the base pairs—the rungs of the double helix—opaque and suddenly visible under an electron microscope. This wasn't as easy as it sounded, and Glover commendably presented pictures of failed experiments showing globs of tangled, unreadable DNA. But then he showed a photograph that vaguely resembled the tire tracks of the lunar lander.[2]

These looked like the actual rungs of the double helix, the iconic helical corkscrew stretched and flattened into a straight ladder, potentially stretching for thousands of bases.

A similar effort, backed by George Church, was progressing at a Silicone Valley start-up, Halcyon Molecular, developing "a fearsomely powerful tool for unraveling secrets of biology, disease, and aging." Company president Luke Nosek, a cofounder of PayPal, bragged that Halcyon would ultimately sequence complete human genomes in ten minutes for $100 or less, and at an astonishing (99.9999 percent) accuracy.[3] The scientific cofounders, the Andregg brothers, stretched out single strands of DNA on a sheet of carbon, attaching platinum atoms to the G bases and osmium to the T bases. (The same procedure could be repeated for the complementary strand to deduce the positions of the As and Cs.) The method could deliver accurate complete genomes with single reads up to 200,000 bases.

There were many other innovative approaches. One, developed by an outfit in upstate New York called Reveo, created submicroscopic devices known as nano-knife edges to penetrate the crevices between the DNA bases to read out the sequence. In 2009, Boston's Intelligent Biosystems reported that its PinPoint sequencer would ultimately sequence a human genome for less than $200.[4] The big guns, such as GE Healthcare and IBM, were taking interest too. A chance hallway conversation between two IBM scientists at Big Blue's research center just outside New York City spawned a project to develop a semiconductor sequencer based on synthetic nanopores.[5] Using a multilayered design, IBM could ratchet a single strand of DNA through an artificial channel one base at a time. Detecting the base could be done by measuring current or capacitance, although IBM was still a long way from turning theory into reality. And in 2010, an MIT start-up called GnuBio outlined a microdroplet method that could one day deliver a $30 genome.

But of all the firms bidding for a shot at scientific superstardom, three warranted particular attention. Oxford Nanopore aimed to continue what Solexa had started and demonstrate that the United Kingdom could still be a force in DNA science. Jonathan Rothberg returned with his exciting company, Ion Torrent Systems. But nobody had amassed the war chest of Pacific Biosciences, which could literally eavesdrop on DNA synthesis.

In February 2008, I flew down to Fort Myers, Florida, accompanied by a planeload of baseball aficionados looking forward to Boston Red Sox spring training. But my journey continued an hour south to Marco Island for the annual gathering of 700 of the world's leading genomics researchers. The major sponsors were the usual suspects—Roche, ABI, and Illumina—all offering free food, drink, and entertainment. On opening night, a band dressed as pirates flambéed such classics as "Free Bird." The following night, we admired a fairly impressive beachside fireworks display, with the pyrotechnics sponsored by San Francisco's Pacific Biosciences, a mysterious newcomer to the scene. All was revealed less than forty-eight hours later when PacBio's founder, Stephen Turner, confidently took to the podium to deliver the final talk of the meeting. His presentation was so dazzling that I wondered why his colleagues had bothered with the fireworks at all.

Flicking through the *New York Times* earlier that Saturday morning, I stumbled on the first mainstream report on the company.[6] One comment in particular caught my eye: the CEO, Hugh Martin, said PacBio was going to win the X Prize. Fortunately I had an appointment to meet Turner and Martin that same morning.[7]

The decision to come out of stealth mode was taken three months earlier, when Turner's team managed to sequence a fifty-base stretch of a single molecule of DNA, convincing the PacBio directors to approve full-scale development. Lead investor Kleiner Perkins had tapped Martin, a former telecom entrepreneur, to be the CEO and make good on the commercial potential of Turner's technology.[8] Martin's objective was to build a public independent company rather than become part of someone else. "We have at least a fifty-fifty shot," he said.

Dressed conservatively in sports jacket and slacks, Turner looks barely old enough to be out of graduate school, although he earned his PhD at Cornell almost ten years earlier. He stayed on to do his postdoc, waiting for his wife to earn her degree while he honed ideas of applying cutting-edge nanotechnology and biochemistry to observe DNA synthesis in real time. In 2001, he launched his company, but his first funding deal collapsed in the ruins of September 11. He limped along with support from the NIH and scrounged office space from another Cornell start-up. Finally, Bill Ericson of Mohr, Davidow Ventures agreed to pump in $5.5 million and persuaded Turner to move to California. The company was originally called Nanofluidics, but Turner had concluded the *nano-* prefix was no longer a mark of a quality company. At an after-hours naming party over pizza, Martin and Turner hatched the new name.[9] "We wanted the name to convey a sense of quiet strength, something that would be comforting to a person in a clinical setting," said Turner, already projecting years ahead and not wanting to alarm patients with Crichtonesque visions of swarming nanobots. Turner's mission was nothing short of the use of whole-genome sequence data in routine medical care for people.

As the genome project was reaching a crescendo, Turner took a fresh look at DNA polymerase—the enzymatic staple of Sanger sequencing. He believed it was a seriously underappreciated molecule, capable in nature of sequencing nearly 1,000 bases per second. "Its read length is not 800 bases but 800,000 bases! It's extraordinarily frugal, [consuming] less than a femtogram of material. It almost never makes mistakes and it's physically very small. So how do we take advantage of it?"

Turner's idea, developed with his Cornell friend Jonas Korlach, was to eavesdrop on single molecules of DNA polymerase synthesizing virgin DNA in real time. He would attach fluorescent tags to the four constituent bases in solution and record every instance that a new base was snagged by the polymerase into the growing DNA strand. Turner dubbed the concept "single molecule, real time" (SMRT) DNA sequencing. To eavesdrop on DNA synthesis, Turner needed to train a spotlight onto a single isolated enzyme. The ingenious trick was to borrow a principle from a standard kitchen appliance. When you peer through a microwave oven door, you're actually looking through a meshed screen with holes too small to permit the microwaves to pass through. Turner applied the same concept on a nanoscale, reducing the wavelength from microwaves to visible light of 500 nanometers and shrinking the holes to just a few tens of nanometers in diameter. Illuminating the holes from underneath with a laser, only a tiny zone of light (evanescence) passes through—just enough to bathe the polymerase sitting at the bottom of a tiny well in light. The tagged nucleotides remain in the dark, coming into the light only when one is grabbed by the enzyme before being stitched into the new DNA strand.

The concept is called a zero mode waveguide (ZMW) and graced the cover of *Science* in January 2003.[10] It consists of a thin layer of aluminum sitting on a silica slide. A grid of tiny wells—each well holds just twenty zeptoliters—is etched into the aluminum, into which is dropped a solitary molecule of DNA polymerase. The camera barely registers the fluorescently tagged bases diffusing in and out of the ZMW on a microsecond timescale. But when the enzyme snares the appropriate base, the two snug for near eternity (actually only a few milliseconds). The camera detects the fluorescing dye before the enzyme ratchets along the DNA to the next position.[11] In nature, the polymerase moves too fast for current imaging technology to detect. "Even if we could go 1,000 bases per second, we would not want to," Turner said.[12] (The current setup is timed to a rate of a few bases per second.)

The result was not perfect, Turner admitted, but the accuracy was high. More important, this was a completely new way of looking at sequencing: watching a polymerase work in real time as DNA synthesis happens. I was itching to see some real data, and Turner gladly obliged. He launched a video showing a grainy, magnified grid of 1,000 ZMWs that would fit onto a pinkie fingernail. When Turner added the final ingredient, the ZMWs burst into a riot of sparkling lights—like flashbulbs at the first pitch of the World Series—as the polymerases jumped into action. And this, he reminded me, was in slow motion (the flashes in real time would be too fast to distinguish). Martin was beaming: "No one's ever seen 1,000 polymerases making DNA before in real time. It's pretty remarkable." I would have agreed, if I wasn't shaking my head in amazed disbelief.

Turner argued that his approach was better than that of the competition in every respect: time, cost, and data size. "Instead of being days, it is minutes for a run. Instead of being thousands of dollars, it's hundreds of dollars for a run. Instead of being terabytes, it's gigabytes for the data size. Instead of being hours per base, it's bases per second. It's 30,000 times faster than the fundamental sequencing unit of platforms available today." In another neat trick, Turner could circularize the DNA molecules being sequenced, so the sequence could be repeated and checked multiple times. Here, then, is Turner's vision of SMRT sequencing: single molecules sequenced at superspeeds, a sixth sense for medicine. Within a few years, Turner was shooting for speeds of 100 gigabases per hour. "The ramification of that is: a draft copy of the human genome in 2.5 minutes, and with [appropriate depth of] coverage, 15 minutes to complete a human genome! This is an extraordinary number but—if you allow me to hark back to the mission of the company—this is what is required to get human genome sequencing into routine medical practice."

Finally, I asked Martin why he would brag to the press that his company was going to win the X Prize. "You may be hearing a little bit of pent-up demand from three-and-a-half years of not saying a single thing, but we now know it will work," Martin said. Turner's "fifteen-minute genome" pledge made me recall the overenthusiastic claims of Eugene Chan at U.S. Genomics six years earlier. "I know Eugene," Turner interjected. "The mistake they made is that they started their company rollout before they had the certainty that the technology would work. We wanted to be sure it was going to work before we said anything."

A few hours later, anticipation building as Turner walked on stage for the closing talk of the conference, 454's Michael Egholm leaned across to a colleague and whispered, "This better be good." Turner delivered the same eloquent story he'd rehearsed for me that morning. He closed by thanking his coworkers. "Their autographs are literally on the back of this presentation," said Turner, to enthusiastic applause. "Wow, how cool was that?!" purred chairperson Elaine Mardis, chief of The Genome Center at Washington University.[13] That evening, as scientists guzzled cocktails amid the balmy Gulf Coast breezes, it was clear Turner was the hit of the meeting. One competitor told me, "We better make hay while the sun shines." Even Egholm later admitted some of Turner's data were "mind boggling . . . worth the cover of *Science* and *Nature*."

When Turner returned to the Marco Island stage twelve months later, PacBio had just published its first single-molecule sequencing results.[14] The early sequence data contained errors, and the tempo of the process was highly irregular. But these were old data and would be little more than a nostalgic timepiece by the time PacBio eventually unveiled its prototype. Turner still wowed the audience. Early on, he launched a movie of a real DNA sequence trace, which continued running hypnotically at the foot of his slides, like a stock ticker, for the entire talk. "I do hope that some of you will watch the rest of the talk," Turner said with a smile. The crawler script was written by a member of the software engineering team headed by Scott Helgesen, the ex-454 star who had been lured back to the sequencing fold by PacBio's technology and the California sunshine. Instead of impressing the PacBio interviewers with his *Men in Black* exploits, this time it was Helgesen's turn to be blown away by the new challenge of real-time data processing.

It wasn't just the scientific community that was impressed. *Fortune* called PacBio Silicon Valley's "hottest start-up," and predicted it had a shot at becoming "the Google of health care."[15] By the end of 2009, PacBio had raised an extraordinary $260 million in venture funding, much of it during a deep recession. The number of employees had climbed to 300, including new chief science officer Eric Schadt, who had come from Merck.[16] Martin said the prototype would feature a price of $100 per run, but regular upgrades would ensure the amount of sequence for that price would grow dramatically. As for the commercial release, the instrument would sequence three bases per second for at least fifteen minutes, with a complete experiment taking less than twelve hours.[17] As Stanford chemistry Nobel laureate Roger Kornberg said, "If they succeed, they own the business."[18]

At the 2010 Marco Island meeting, Martin and Turner shipped one of their prototype machines across the country and parked it in a hotel suite under tight security. The first thing that struck me about the breathtaking PacBio RS was that for an instrument sequencing single molecules of DNA, it's awfully big. The imposing floor-standing machine weighs about 1,900 pounds and is 6½ feet wide. It is the Rolls-Royce of sequencers. "It's packed," said Martin, somewhat defensively, with robotic arms, high-speed cameras, and optics. "You need a pretty heavy-duty floor," he conceded. That didn't seem to bother ten customers desperate to get their hands on what Martin called the first "third-generation" sequencer, and they were even willing to pay full price—$695,000—to get one. "It's really the world's most powerful real time, single-molecule microscope," said Martin, and a quantum leap for the field.

During his latest talk before another standing-room-only crowd, Turner presented one sequence trace of more than 10,000 bases, which was truly impressive. The machine is loaded with 96 SMRT cells, each holding 80,000 ZMWs.[19] There was also a clever strobe method to conserve the polymerase enzyme (which is prone to damage) and push read lengths out longer.[20] At launch, the machine will be sequencing runs of 1,000 to 1,250 bases, with a fraction of the reads up to 5,000 bases. The goal was to read lengths of 40,000 bases and more than 100,000 bases under strobe conditions. "There's no one else out there in third gen," said Martin. But what was Martin's definition of third-generation sequencing? I anticipated some conceptual definition involving real-time sequence, but he said it was "everything that second gen is—throughput, cost per base, etc.—with the addition of very long read lengths, extremely low reagent or consumable cost, and very fast run times."[21]

It had taken Turner's team two years to progress from a fifty-base sequence proof-of-principle to manufacturing an instrument, which gave him confidence that they had two to three years advantage over the competition. But to reach the fifteen-minute genome, PacBio would need to be tracking 1 million ZMWs in unison, not 80,000, and that would require a new instrument, scheduled for 2014. And beyond that, Turner even hinted at the possibility of *millions* of ZMWs housed in a handheld device, which would truly revolutionize health care.

A few hours later, we gathered on the Marco Island beach for more PacBio pyrotechnics. I congratulated Turner not only on producing his first sequencer, but also on the lost art of dual-slide projection, a new gesture during his characteristically slick presentation. But showing an impressive, almost compulsive attention to detail, he'd already read my Tweet on the subject and all the other reviews besides.

One of the problems with a 1,900-pound sequencer is that it doesn't lend itself to easy installation in a clinic or doctor's office. Imagine instead a third-generation sequencer as small as a DVD player, providing the same output without lasers, fancy digital cameras, or stabilizing slabs of granite. In the city of dreaming spires, Oxford Nanopore Technologies, by contrast, is building a sequencer that dispenses with light and cameras. "It's label free, reagent free, electrical single-molecule sensing," says CEO Gordon Sanghera. Instead of imaging and pictures, the sequence is determined by reading variations in electric current. What could be simpler?

Oxford Nanopore was cofounded by Hagan Bayley, a professor of chemical biology at Oxford University.[22] After spending nearly three decades in the United States, Bayley was lured back to Oxford in part by the university's stunning new $120 million chemistry department, which replaced the venerable Dyson Perrins building.[23] "The geographers are in there now, breathing in all the mercury fumes," he chuckled in his spacious new office.[24] "There are quite a few dotty geographers around." Bayley has thick silver hair, a beard, round-rimmed glasses, and speaks in the laconic baritone of a late-night radio DJ. If he has an entrepreneurial streak, it's not obvious. "I feel very fortunate to be a professor despite all the terrible things they do to us day to day," he said. "I enjoy that life and get paid enough and stuff, but at the same time, if you develop something, it's nice to see it get used."

In 1996, Bayley's group deduced the atomic structure of a bacterial nanopore called alpha-hemolysin, a sort of molecular hole puncher that looks like a hollowed-out drawing pin.[25] Bayley reviewed many applications of these "doors in cells" in an article for *Scientific American*,[26] but one of the most promising was demonstrated by Harvard's Daniel Branton and David Deamer, who showed that DNA molecules could snake through the nanopore, which was just a billionth of a meter wide.[27] But Bayley was more interested in building medical devices, from sensing flu viruses to delivering drugs to cancer cells. In one paper, he sketched a *Fantastic Voyage*–style submarine as a kind of futuristic nanobot for detecting and delivering drugs in the human body. In fact, Bayley stayed away from DNA sequencing as a professional courtesy to Branton, making progress in chemically modifying the hemolysin to use it as a single-molecule sensor for drugs, chemicals, and metal ions. The trick was to insert cyclodextrin, a small, rigid, sugar-like molecule that acted as a grommet that could slow down molecules passing through the channel.

In 2005, Bayley met with members of the IP Group, which had helped fund Oxford's new chemistry building, and together they formed a spin-off company to commercialize nanopore technology. Spearheading the deal and joining Oxford Nanopore as chairman was IP Group founder Dave Norwood, a brilliant investor and former captain of the English chess team.[28] His colleague James "Spike" Willcocks, Oxford Nanopore's head of business development, is one of the top court tennis players in the world.

After winning a tidy $4 million grant from the NIH, Bayley felt it was time to jump back into DNA sequencing. He said, "By that time, I think those other guys at Harvard had had ten years to do something, so the statute of limitations had expired." In 2006, Bayley's team showed that by measuring dips in the current passing across a nanopore, they could discriminate among the four different bases of DNA as they slipped through the channel.[29] Oxford Nanopore now had its commercial vision—a sequencing chemistry that did not require any fussy fluorescence or pricey hardware.

Building a machine around a protein sounds risky, but Bayley insisted the natural bacterial protein pores were extremely stable. "Hemolysin will operate almost in boiling water," he said. "The device could be very small, so what you see in the movie *GATTACA* is not completely unreasonable. Our early models look very similar!" Bayley imagined a compact, high-throughput sequencer that could sequence thousands of bases in a single strand of DNA, with hundreds of thousands of reactions running in parallel. "It'll tell us a huge amount about what it means to be human. . . . It's very possible that everyone will be sequenced at birth."

I first met CEO Gordon Sanghera and Willcocks in San Diego in spring 2008. Neither had any intention of speaking publicly about Oxford's plans yet. Sanghera was biding his time but promised that when Oxford was ready to emerge, it wouldn't be as part of the next-gen movement. "It's a revolution rather than evolution," he said.[30] Building a successful British biotech company was almost as important. "The last company I worked for, MediSense, was Hoovered [vacuumed] up. Solexa was Hoovered up. We've all bought into the vision that we're going to make a British company work." Sanghera had already raised $35 million, much from hedge fund managers interested in a big win.

The Oxford Nanopore system works by inserting the nanopore protein into an artificial cell membrane separating two chambers. When a small voltage is applied across the chambers, current flows. As each base tumbles through the pore, past the cyclodextrin washer, it impedes the current by a fixed but discrete amount, dependent on subtle differences in size and shape of the As, Cs, Ts, and Gs. "You literally just read off the current," Sanghera explained. For example, "If you're between 19 and 21 milliamps, it'll be a C. It's a direct electrical read."

To move to the all-important step of actually sequencing DNA, Oxford borrowed an idea originally championed by Kevin Ulmer at SEQ in the 1990s. It could best be summed up in the highly technical phrase: "Chop it and drop it." The idea was to attach an exonuclease to the lip of the nanopore: the enzyme would snip the end of an incoming strand of DNA like a PacMan, one base at a time.[31] Each newly freed base then tumbles into the nanopore for its identity to be revealed by the transient blockage in current. "We're targeting ten to thirty bases per second, which is important," said Sanghera. "It all has to line up. It's like the London buses: they can't just all turn up at once because they'll get mixed; you won't know which bus is which!" Willcocks added that the bases would file in order through the pore because at that nanoscopic scale, the current flowing through the pore was equivalent to a bolt of lightning.

In principle, Oxford could squeeze thousands of nanopores onto a chip. "Each hole is individually addressable, a worker bee," said Sanghera. A 10,000-nanopore chip sequencing at thirty bases per second and eightfold coverage could read off a whole genome in one day in a box the size of a DVD player. "With the silicon semiconductor industry, this type of chip is very simple," added Willcocks. "To go to hundreds or thousands [of wells] is pretty simple. This pitch is massive compared to Intel. We just need to get it working."

Oxford wasn't the only nanopore biotech company, but Sanghera didn't appear the least bit concerned.[32] When the time was right, there would be no gimmicks or Floridian fireworks. "We'll just give our [debut] presentation in a very dry, British way. We might open up with *Monty Python*'s, 'And now for something completely different!' That'll be as close as we get. We want to come over as a very reputable science company with amazing investors." I had my doubts that a nude John Cleese playing a theater organ was the classy image Sanghera was striving for, but let it pass. "We are *the* nanopore company," Sanghera insisted. "Electrical, single-molecule sensing on an instrument the size of a laptop is going to transform a lot of areas, not just DNA sequencing."

Several months later, touring the Oxford labs in safety goggles and ill-fitting lab coat, I met Marion Crawford, an expert in analyzing the single purified nanopores despite her youthful appearance. Above her bench was a monitor tracing a series of jagged mountains and troughs as current passed through a single nanopore. Running a finger along the display, she effortlessly read out "T-G-T-A-C . . . ," identifying each base by its respective dip in the current. I had to take her word for it, but she'd clearly seen enough of these to convince me. A young scientist named James Clarke was putting the finishing touches on an important paper describing this impressive base discrimination, but had just realized that the working title somehow omitted the word *nanopore*. After two years, Clarke had figured out exactly where to place the cyclodextrin ring inside the nanopore to identify the passing bases.[33] As an added bonus, the setup could also distinguish bases that have been methylated, that is, tagged with a small chemical group widely thought to be a crucial mechanism in regulating gene activity (a field known as epigenomics). In charge of building and miniaturizing the nanopore chip was vice president of engineering Simon Wells, who showed me a series of thumbnail-sized chips, each containing a denser cluster of wells. On this day, he was at 128 wells on a single array, and looking for more.

But of all Oxford's recruits, the two biggest arrived in June 2008. Clive Brown and John Milton had already tasted success at Solexa and could take much of the credit for building the platform producing most of the world's sequence data. After leaving Solexa, Brown had been working at the Wellcome Trust Sanger Institute, but on visiting Oxford he was sold after thirty minutes. "They wanted to give me the job, but I said, 'Look, you've got to employ John. You need decent industrial chemistry.' " Milton said he understood the technology requirements from end to end now and was equally enthused. "The speed, the chemistry, the electronics, there are certain boxes that need to be ticked, and Oxford Nanopore ticked all the boxes."

The pair enjoy a comfortable camaraderie, needling each other relentlessly. Milton joked that Brown views chemistry like the elementary ball-and-stick models he used as a teenager. Brown rebutted that Milton considers software as nothing more than glorified spell checking. As the lead chemist, Milton was sold on the remarkable robustness of alpha-hemolysin, an essential prerequisite when Oxford started to ship its sequencing instruments. "Thankfully, a sample of alpha-hemolysin several years old is as good as the day it was made," he said.

Brown's challenge was to develop the software to read the sequence. He wanted a benchtop analysis pipeline that scientists and clinicians, not computer geeks, could operate, even remotely if necessary. The specs of the box could be anywhere from 40 to 100 million bases of sequence per second per box, or 3 to 5 terabases per day, with up to twenty boxes stacked together.[34] Illumina saw enough potential to ink an $18 million deal for the marketing and distribution rights for the exonuclease sequencing technology, but Oxford was also contemplating other applications eventually, with the goal of becoming the leader in diagnostics technology. One feasibility study was with the U.S. Department of Defense for an airport security drug detector. "We've got some heroin in the office fridge," said communications director Zoe McDougall cheerily.

While Oxford's executives couldn't openly share the specifics of their latest progress, Milton was happy to dish on his rivals.[35] Helicos, maker of the first single-molecule sequencing instrument, had reached a similar point to Solexa six years ago. "It's comatose, not Helicos," said Milton. PacBio had "some quite flashy stuff, but there's a huge difference between sequencing ΦX and the human genome. We don't see them as [third-gen]. If it's fluorescence and it's the size of a fridge and it's optics, it's current generation." Nor was Milton threatened by the Complete Genomics service model. The only way it could break even, he said, was basically to do no sequencing. "If you're going to pay American technicians on American salaries and American rents, the numbers don't add up," he said. "The way they do this calculation to do 'the $1,000 genome,' it's just ridiculous. They just put so many numbers to the side, but those numbers exist and have to be paid." Of course those extra computing and storage costs, overheads, and other costs would have to be paid by Oxford Nanopore as well.

Milton said the nanopore sequencers would not only be faster but much smaller than those of the competition. "I'm not saying they're going to be *Star Trek*–like, but they're going to be like a video recorder," he said, miming the dimensions. "You're going to be able to stack them. In our model, we've got a stack of these things, up to the size of a file cabinet. One of those has got the same power as about four genome centers." What is more, "There's no reagents—just salt water!" Oxford wasn't shooting for a $5,000 genome or even a $1,000 genome. Based on the way their competitors were calculating the costs of genome sequencing, Milton declared, "We can do a nought-dollar genome!"

Oxford Nanopore wasn't the only firm touting a miniature electrical platform for DNA sequencing. In February 2010, the irrepressible Jonathan Rothberg finally unveiled his new company, Ion Torrent Systems. One day after Pacific Biosciences debuted its $695,000 machine at Marco Island—one rival cruelly likened the 1,900-pound behemoth to Lenin's tomb—Rothberg proudly showed off his svelte new semiconductor sequencer—a snip at just $50,000.[36]

Just as Rothberg credited the idea for 454 to the birth of his son, Noah, the genesis of Ion Torrent also owes a debt to his precocious child. After the Watson genome presentation in May 2007, the newly unemployed Rothberg went home and proudly told Noah, then eight years old, that he'd just presented Jim Watson with his DNA sequence. "What does that mean?" Noah asked.

"I scanned his genome," said Rothberg.

"Why don't you make something that will scan people's minds, so you know what people are thinking?" Noah replied.

While Rothberg couldn't figure out how to sense the brain's neurotransmitters, he began to ponder a semiconductor device for monitoring the chemical process of DNA synthesis. This time, he would focus on a by-product of the reaction—the hydrogen ion released during every base incorporation, which would increase the solution's acidity. What Rothberg needed was the world's smallest pH meter. In barely two years, Rothberg conceived and built the first "personal genome machine," so named not because it could sequence a personal human genome (at least not yet), but because it was designed to be the staple for almost any researcher around the world. To make the point, during his Marco Island presentation, two colleagues carried in the machine, which had been sequencing DNA in a nearby hotel suite. It was about the size of a desktop printer, accented in blues and oranges like a children's toy.

"It's super cheap, a tenth as much as any next-gen on the market," he said. To Rothberg, this technology was so revolutionary, it wasn't even next-gen sequencing. It deserved a new name: post-light, semiconductor sequencing. If Sanger sequencing was the mainframe and next-gen sequencing was the minicomputer, Rothberg was developing the sequencing equivalent of the personal computer. In next-gen, the box was the machine. Now the chip was the machine.

The Ion Torrent sequencing method bore some similarity to the 454 method Rothberg had developed almost a decade earlier. At the heart of the instrument was a typical semiconductor chip, made in the same foundries that make the chip for the iPhone. The DNA sat in series of 1.4 million microwells. Directly underneath was a sensor that detected the pH and converted it into a voltage every time a base was incorporated.[37] "It's the simplest sequencing in the world—one well, one sensor, one read." Neither the enzyme nor the bases were modified, so the DNA synthesis was totally natural, unlike that of much of the competition. Rothberg said he wasn't going to fight 1 billion years of enzyme evolution. After all, he said, the polymerase has already been modified by God.[38]

Rothberg sees his new baby being used for all kinds of research and clinical applications around the world. "You can sequence on the back of a donkey if you want to," he said. Evidence of his international intentions was stenciled on the front of the machine, above the four tubes containing the nucleotide solutions. Instead of labeling the receptacles A, C, T, and G, Rothberg had bizarrely chosen four symbols:[38]

○ × □ +

"Our users are not going to know what As, Cs, Ts, and Gs are," Rothberg told me as we previewed the machine in the hotel suite. There was just one problem: he couldn't remember what the symbols stood for. "Oh, this is embarrassing," he laughed, frantically beckoning his colleagues to help him out. "Jonathan doesn't know the symbols!" he shouted. I should have asked the obvious question: Would the actual DNA sequence also be printed out in his elementary new genetic alphabet?

The first Ion Torrent machine could read about 100 to 200 bases in an hour or two, so it was not an immediate threat to the capacity of Illumina and Life Technologies. But Rothberg stressed he wasn't competing with the establishment. "This chip is not for whole-genome sequencing of humans," he insisted, at least not yet. It was for sequencing bacteria, measuring gene expression, studying metagenomics, and for a host of future apps that, just as when the iPhone launched, no one had even thought of. Eventually he expected to achieve read lengths of 1,000 bases and up, from millions of wells on a chip—Rothberg was already planning a chip with 7 million wells. The more he squeezed onto the chip, the greater the performance and the cheaper the cost, just like the relentless improvements in computer power enshrined in Moore's law. And because no light or imaging was involved, the demands on data storage were a fraction of the competition. Ion Torrent's computer processing came in the form of a regular laptop.

It was semiconductor sequencing: Gordon Moore meets Jim Watson, with a splash of Steve Jobs for good measure—three of Rothberg's biggest heroes. "I've got pictures of my son with Gordon Moore," he said proudly. "Well, wouldn't you?"

CHAPTER 12

Personalized Response

On June 15, 2005, in a dimly lit Holiday Inn ballroom just outside Washington, D.C., renowned cardiologist Steven Nissen and fellow members of a Food and Drug Administration (FDA) advisory panel listened intently as executives from a small drug company, NitroMed, described the remarkable success of its heart failure drug, BiDil. They were followed to the microphone by a procession of politicians, activists, ethicists, and legal scholars anxious to air their views both for and against the controversial drug. It wasn't just that BiDil's very existence hinged on the shaky supposition that it worked better in blacks than whites, but that the U.S. government appeared to sanction the idea of race-based therapy. BiDil was billed as the world's first ethnic medicine, one welcomed by many cardiologists as a potentially life-saving treatment for a chronically underserved community. But ethicists charged that the drug was nothing short of a government-sanctioned reification of race.

We have already seen numerous examples of how knowledge of one's personal genome can provide compelling insights into genetic risks and disease diagnoses. But what can it reveal about the likely success and potential dangers of drugs we might be prescribed? The tailoring of drugs to specific target populations has been a tantalizing prospect for drug companies for years. While no drug company in its right mind would wish to abandon a blockbuster drug pulling in at least $1 billion annually, drugs designed to work with specific subsets of patients harboring known mutations or shuffled genes can pay off with faster development times, longer patent lives, and different pricing models. The BiDil saga showed what can happen when another factor, in this case skin color, is used as a tenuous substitute for robust genetic information.

BiDil was the brainchild of Jay Cohn, a University of Minnesota professor and the founder of the Heart Failure Society of America. Heart failure and its chief symptom, hypertension, affect more than 5 million Americans, with some 700,000 new cases diagnosed each year. It disproportionately affects African Americans because of a variety of socioeconomic and environmental factors: access to medical care, disease management, education, and poverty. In the 1970s, Cohn devised a combination of two generic drugs (later dubbed BiDil) to treat heart failure by relaxing blood vessels through the generation of a natural signaling biochemical, nitric oxide (NO).[1] Cohn led a major clinical trial, called V-HeFT (the Vasodilator Heart Failure Trial), involving more than 600 patients, including 180 African Americans, but BiDil fared less well than an ACE inhibitor.[2] Cohn was later granted a patent on the BiDil combination in 1989, but there was no mention of possible racial differences, and years later, in 1997, the FDA turned down his first attempt to get BiDil approved.

With his patent running out, Cohn reassessed the V-HeFT data, and by stratifying the data according to the self-identified racial background of the volunteers, he uncovered a dramatic difference between what he called "the blacks and the whites."[3] With some trepidation, he finally published his new theory of heart disease and race in 1999 while bracing for the controversy.[4] Hypertension, he argued, had different physiological origins in blacks and whites and so should be treated differently. He hoped that BiDil might prove as effective for treating African Americans as ACE inhibitors in whites, and he proposed new trials involving large numbers of black patients. The FDA agreed, even though the putative benefit of BiDil was based on just forty-nine African-American patients.[5] Cohn assigned the rights to BiDil to NitroMed and filed a new patent application for the specific treatment of heart failure "in non-Caucasian patients, preferably black patients." The U.S. Patent Office controversially deemed this to be a "nonobvious" extension of Cohn's original patent, and in 2002 it approved the application, thus granting NitroMed exclusive rights to BiDil until 2020—effectively a thirteen-year patent extension.

With support from the Association of Black Cardiologists (ABC) and the Congressional Black Caucus, NitroMed began enrolling more than 1,000 African-American patients in the African-American Heart Failure Trial (A-HeFT), by far the largest number of African Americans in any heart trial in history. The results impressed the FDA so much it suspended the trial in July 2004. Only thirty-two patients among the group taking BiDil had died, compared to fifty-four in the placebo group—a dramatic 43 percent reduction in mortality.[6] NitroMed's stock price soared as cardiologists praised such spectacular results from a tightly designed trial. Law professor Jonathan Kahn concurred—up to a point. "A-HeFT was a very innovative, single-race trial," he said, but the innovation was not so much that it was single race, but that the single race wasn't white but African American. "What it did show, quite strikingly, is that BiDil works—and that's a good thing. What it didn't show is a racial difference."[7] New Orleans physician Keith Ferdinand, the chief science officer of the ABC, staunchly defended the design and results of A-HeFT: "The reason you do a trial in a subgroup is to show benefit in that subgroup, not to show it won't work in other groups. This ain't no Tuskegee!"[8]

Nevertheless, although the infamous Tuskegee syphilis study was not mentioned explicitly, its presence at the FDA hearing "was as palpable as a haint in a Southern gothic town," recalled Susan Reverby, a Wellesley College historian of medicine.[9] Congresswoman Donna Christensen implored Nissen's panel to begin to reverse 400 years of health disparities and discrimination against

African Americans. A patient testified she felt fabulous thanks to her "strong faith in God and a little pill called BiDil." On the flip side, anthropologists pleaded with the panel not to sanction "patenting blackness" or approve a "black drug." Kahn favored approving BiDil, but for use in the general population. Just as drugs that are approved based on trials conducted predominantly in white patients are not designated as *white* drugs, neither should BiDil be designated as exclusively a drug for African Americans. He later wrote, "Medical researchers may use race as a surrogate to get at biology in drug development, but corporations are using biology as a surrogate to get at race in drug marketing."[10]

Nissen was singularly unimpressed by such concerns. "There were a lot of ethicists, almost all white, who said approving BiDil is unethical," he told me later.[11] "But every African-American group said, 'Finally, someone is developing drugs for African Americans.' " Nissen found the evidence for BiDil saving lives compelling. Patient selection by race was simply a pragmatic stop-gap until doctors could practice a more refined form of personalized medicine. Nissen's panel voted unanimously to recommend FDA approval, a decision the FDA quickly ratified, to the dismay of geneticists such as Duke University's David Goldstein, who said, "Race for prescription is only an interim solution to carry us through a period of ignorance until we find the underlying causes."[12]

The Human Genome Project (and the subsequent HapMap project) demonstrated convincingly that humans do not conveniently congregate into a few tidy population silos, but rather lie on a seamless continuum of genetic variation between individuals within and between different population groups. New data show that one of the oldest human populations, Kalahari Bushmen, are genetically more diverse than, say, a European and an Asian.[13] We are all Africans in a sense (all our ancestors moved out of Africa at some point in the past 100,000 years).[14] Moreover, slaves were brought to America not from one region but from the breadth of continental Africa—hardly a simple or common ancestry. When they arrived, the gene pool was further muddied by intermixing with Europeans—

a phenomenon known as "admixture." The highly publicized cases of Thomas Jefferson and his slave Sally Hemmings, and more recently Henry Louis "Skip" Gates, are prominent examples.[15]

Despite its success, A-HeFT did not answer one critical question: Did BiDil actually work better in blacks than whites? Cohn surmised that blacks had a more dramatic response than whites to the drug based on a convoluted rationale: "Black people have less NO activity, their blood vessels are more constricted, they have a higher incidence of hypertension, and higher risk of vascular disease. All that makes this population somewhat phenotypically different—and maybe genotypically different."[16]

So what was it? Nature or nurture? Genes or environment? Goldstein and fellow Duke geneticist Hunt Willard posed the question in a *Boston Globe* op-ed:

> Is it because African-Americans are exposed to more lead, pollutants and pesticides than Americans of European origin? Or does the answer lie more with the 10 to 15 percent of our genetic variation that is associated with geographic or racial groupings—the hundreds of thousands of bits of genetic code that are characteristically different between, say, Africans, Europeans, and Asians.[17]

Despite his conviction that his drug worked best in blacks, Cohn didn't exactly want to exclude whites from the potential benefits of BiDil. "It would be amazing to me if the drug were not effective in white people," he said. "There's nothing that different between whites and blacks. They have the same disease process, so it's likely it works in everyone." Then again, had his original trial confirmed that suspicion, Cohn would have had no grounds to extend BiDil's patent. The debate over BiDil's putative predilection for black biochemistry could be settled by a new clinical trial comparing different ethnic populations, but that wasn't in the cards. Even if it was deemed ethical to administer a placebo to black heart failure patients, NitroMed didn't have the resources to do a big $100 million trial comparing subpopulations of blacks, whites, and Asians. "What, you want to do another trial where another couple of thousand people die?" said Ferdinand incredulously.

NitroMed fiercely defended its marketing on the basis of skin color. Aside from genetic predisposition, environmental factors such as smoking, obesity, diabetes, or lack of access to the health care system because the patient is poor and arrives at a medical center when disease has evolved too far to be reversed, were contributory factors.[18] Analysts had predicted annual BiDil revenues around $100 million, but in fact they barely topped $10 million due to a perfect storm of bad things: pricing, lack of reimbursement, promotion, and education. While a year's supply of BiDil could cost upward of $4,000, on top of other drugs that heart failure patients take, physicians could prescribe the generic alternatives for pennies on the dollar. And despite the influential support of black organizations in approving BiDil, suspicion lingered among some African Americans that a racially targeted drug must be wrong. The memory of Tuskegee still resonated. Recruiting African-American patients for A-HeFT, Boston Medical Center's Flora Sam said, "You could see a shield came up. There was this barrier. Patients felt a degree of mistrust."

Cohn said he was comfortable with BiDil's ethnic medicine label and felt he had blazed a path that should lead to more drug trials carried out in more identifiable subgroups rather than megatrials in large populations. He routinely prescribed BiDil for his African-American patients, telling them, "You *have* to be on this drug—I would be negligent were I *not* prescribing this for you." He didn't have the same proof of efficacy in whites, but that didn't stop him from prescribing the drug for them from time to time. NitroMed was trying to have it both ways. At a national cardiology meeting, Reverby noted, "NitroMed covered its booth with pictures of African Americans while handing out that well-known soul food—cappuccino—to passing doctors." NitroMed ran splashy four-page inserts in the top medical journals—"The Buzz Is BiDil"—but it was an uphill struggle. "It comes down to three words: marketing, marketing, marketing," said Nissen. Without the resources of a Big Pharma company, direct-to-consumer ads, and an army of sales reps, one could almost predict the drug wouldn't succeed. NitroMed was eventually sold to Deerfield Management in 2009.

To its credit, NitroMed was embarking on studies to identify genetic factors that might underlie BiDil response before it ran out of money. In time, it might have identified a marker similar to one deCODE had discovered associated with heart attacks. The Icelandic firm identified a particular set of markers associated with a 250 percent increased risk in African Americans.[19] Perhaps these markers were introduced into the African-American gene pool due to European admixture within the past few hundred years—insufficient time for the evolution of other genetic factors that might counteract the increased risk.[20] Such a marker would allow responsive patients to be selected based on their sequence, not their skin color. "It's funny that on a little rock in the North Atlantic, we found a gene variant that has so much impact on the risk of heart disease in a totally different part of the world," joked deCODE's Kari Stefansson. "We may look a little bit different on the surface, but we are all part of the same family."

If the NitroMed saga marks the welcome end of attempts to market drugs based on skin color, are we any closer to choosing drugs that are safe and efficacious based on an individual's genetic profile? With the demise of the pharmaceutical one-size-fits-all blockbuster model amid a slew of safety controversies, plummeting FDA approval rates, and stagnant pipelines,[21] drug companies have belatedly realized that salvation lies in a more personalized philosophy of drug development for the twenty-first century. Chief executives such as Eli Lilly's Sidney Taurel have espoused this new creed: "The right dose of the right drug to the right patient at the right time."

It's a good line, but taken to extremes, it suggests that in the future, we might all be picking our personal drugs as if selecting our favorite candy from the guide in a box of chocolates. Given the dearth of designer drugs on the market in the decade since the HGP, one may be skeptical of such pronouncements, particularly coming from Big Pharma, which has long espoused more of a "me-too" blockbuster mentality. As Eric Hoffman wrote in the *New England Journal*, "Does personalized molecular medicine mean the tailoring of drugs for the individual patient, an approach that evokes images of Bones on *Star Trek* making instantaneous diagnoses with his tricorder, followed by loud pneumatic injections of customized drugs?" Hoffman didn't expect scientists to realize that utopian vision any sooner than they'll be able to develop the warp drive technology that hurtled the *Enterprise* into far-off galaxies.[22]

Nevertheless, there are valid business as well as medical reasons for tailoring drugs to patients. It wasn't just catastrophes like Vioxx, which cost Merck billions of dollars after the heart drug was withdrawn in 2004, that spurred the arrival of personalized medicine. Big Pharma's stagnant pipelines suggest that most of the low-hanging fruit, or drug targets, has been plucked. Nevertheless, the successful development of a niche or tailored drug could allow the developer to charge a premium price, as seen with the high cost of orphan drugs for rare genetic diseases. Genentech's Herceptin owes its success to the ability to diagnose the 25 percent of breast cancer patients most likely to respond to the drug. The approval of Novartis's wonder cancer drug Gleevec in 2001, just nine years after the molecule was synthesized, showed the value of identifying the drug's molecular target (produced by the so-called Philadelphia chromosome rearrangement).[23] But most diseases, even most cancers, don't betray the target in question quite so easily.

Although AstraZeneca's Iressa, approved in 2003, had little effect in most lung cancer patients, 10 percent of patients showed a miraculous response. It turns out that responsive patients harbor mutations that increase the affinity of the drug target, the epidermal growth factor receptor (EGFR). Patients who test positive for EGFR variants are more likely to show a beneficial response, while those who don't are spared false hope and wasted expense. Today many of the newer targeted cancer drugs—Avastin, Tarceva, Erbitux—are administered after genetic testing, while other FDA-approved diagnostic tests, such as Oncotype Dx, look at the expression of a few dozen genes to instruct whether breast cancer patients will benefit from tamoxifen. The goal is to build a sort of genomic cancer classification system, dividing breast cancer and leukemia into various subgroups that define patients who will be treated differently based on telltale molecular signatures, not just the tissue of origin. As for the cancer therapies, they will increasingly involve cocktails of drugs, much like the treatment of HIV, in order to thwart the appearance of mutations that confer resistance. This is already starting to happen thanks to personal-genome sequencing, which is guiding doctors in choosing the right drug in specific instances.

In 1999, Jay "Marty" Tenenbaum, a successful Internet entrepreneur, was diagnosed with metastatic melanoma and given twelve months to live. After desperately researching a range of experimental drug treatments, he credits a failed cancer vaccine with saving his life. Tenenbaum was lucky. He also recognized that patients cannot always wait for conventional drug trials that typically take fifteen years and cost $1 billion or more to develop. So Tenenbaum founded CollabRx to allow patients to tap the world's scientific knowledge and supply resources for treating the long tail of disease. CollabRx focuses on the thousands of drugs that have already passed FDA approval and are considered generally safe. For about $25,000, researchers produce detailed genomic analyses (DNA sequence, gene expression, copy number variants, and so on) of tumor biopsies and generate actionable hypotheses about the molecular mayhem in the cancer and potential interventions.[24] Using that information, CollabRx staff meet with the patient's oncologist to advise on potential drugs or drug combinations, including some drugs still in clinical trials. "This is the way patients with serious diseases are going to be treated," Tenenbaum said. "We're doing it today, and it's only going to get better."[25]

Marco Marra at the British Columbia Cancer Agency in Vancouver, Canada, has reported success in tailoring therapy for cancers based on deep sequencing of the active genes in a patient's tumor. Just as a tailor takes physical measurements of a client before making a suit, oncologists are increasingly measuring their patients' genomic features to gauge which drugs they will—and will not—respond to. In one example, Marra's team sequenced the genome of a tumor that had metastasized from an eighty-year-old man's tongue to his lung. They found that a gene called *RET* was unusually active, while other genes were missing, which probably explained why he wasn't responding to therapy. And so physicians tried an experimental drug, typically used to treat kidney cancers, that blocked the RET protein—with successful results.[26]

Our DNA doesn't just influence the best choice of drugs; in many cases, it has an impact on the activity and overall effectiveness of a plethora of drugs. Naturally occurring genetic variants in enzymes that control the body's ability to metabolize drugs (a field known as pharmacogenetics) play a critical role in governing the safety and effectiveness of many drugs. A Centers for Disease Control report found that nearly 7 percent of all U.S. hospitalizations stem from adverse drug reactions. The $1,000 genome propels the study of genes underlying drug response from bench to bedside. A classic example is warfarin, a widely prescribed blood thinner used for more than half a century to treat patients with stroke and heart attack. Warfarin dosing is notoriously tricky, as patients exhibit a tenfold range in the ability to metabolize the drug. As a result, it is responsible for the second most hospitalizations for adverse drug events, behind only insulin. Give too high a dose to a slow metabolizer, and the drug can cause bleeding and hemorrhage. But administer too little (or an average dose in a fast metabolizer) increases the risk of a blood clot traveling to the lungs, with dire consequences. The FDA now provides dosing ranges on the label based on variants in two genes, *CYP2C9* and *VKORC1*.

Similar insights into the dosage of about 25 percent of all prescription drugs, including depression, come from the FDA-approved AmpliChip, which measures variants in two enzymes involved in the drug metabolism. Uptake has been slow, however, because of the cost (around $500) and lack of insurance reimbursement. Increasingly, the 23 et al. companies are providing information on drug response SNPs for customers. For example, Pathway Genomics provides metabolic information on nine drugs, including warfarin, Plavix, and tamoxifen, as well as providing insight into potential adverse side effects associated with statins.[27]

How will genetic information direct personalized drug treatments in the future? Lee Hood, who invented automated DNA sequencing in the 1980s, espouses a vision in which medicine will—indeed, *must*—change from its current reactive mode, in which doctors treat already sick patients, to a more rational, preventive philosophy of maintaining wellness. Hood calls this "P4 medicine," where the four P's stand for "predictive, preventive, personalized, and participatory."[28] One of Hood's companies, Integrated Diagnostics, aims to market a handheld nano-device with which doctors can measure hundreds of protein biomarkers from blood gained from a simple prick. The results would be transmitted wirelessly to a computer for analysis and a report e-mailed to the patient's doctor within minutes. Another company, Tethys, has developed a test that predicts the likelihood of a patient's developing diabetes, based on an analysis of seven blood proteins, years before the onset of classic symptoms.[29] "In the near future, physicians will collect billions of bytes of information about each individual—genes, blood proteins, cells and historical data," said Hood. "They will use this data to assess whether your cell's biological information-handling circuits have become perturbed by disease, whether from defective genes, exposure to bad things in the environment, or both."[30]

If an individual genome makes medicine personal, detecting protein levels makes it predictive. That knowledge will give doctors insights into health problems years, maybe decades, before they become manifest. And the efforts of firms like PatientsLikeMe and 23andMe are turning medicine into a participatory activity. Doctors themselves will need to become more computer and genetically literate in order to deal with more self-educated patients, who are researching not only their symptoms and family history but their genome sequence as well. Hood sees the developed world exporting P4 medicine to the developing world: "P4 medicine will eventually lead to a universal democratization of health care, bringing to billions the fundamental right of health, unimaginable even a few years ago."

Luke Nosek, the president of Halcyon Molecular, is one of many sequencing entrepreneurs who think they can smash the $1,000 genome barrier and deliver better drugs. Halcyon's technology offers "the first real chance at cures for cancer and other previously intractable diseases," he claimed. "With full sequencing and analysis of millions of genomes, biology can begin to turn into an information science and travel down the path of Moore's law. While we have 10X better computers and video games every ten years, we do not have 10X better cancer cures and we do not really understand what causes the major killer diseases of the first world other than the cop-out term 'aging.' We must change this."[31]

Although the evidence is sketchy, I suspect the molecular therapies of the twenty-first century will prove wondrously effective in combating disease thanks to an expanding arsenal of precision tools available to doctors. It has taken twenty years for gene therapy to recover from serious safety issues and the inflated hype that accompanied its emergence in the 1990s, but successful trials for immune and other diseases bode well. Stem cell research continues apace, buoyed by the sensational discovery that injecting just four gene-activating proteins (transcription factors) can turn almost any developed cell into a malleable stem cell (known as *induced pluripotent stem cells*, or IPS), which can then be reprogrammed into a neuron or cardiac cell, and so on. Many other molecular tricks, including RNA interference and synthetic biology and nanotechnology, offer myriad other therapeutic opportunities.

"Let me suggest something here," said Helicos founder Stan Lapidus as we took our seats at a Chinese restaurant close to Helicos headquarters.[32] I expected Lapidus, sporting his customary professorial bow tie, to suggest some talking points for our lunchtime chat about personalized genomic medicine. I was wrong. "Lightly fried half duck with aromatic flavor," he declared, skimming the menu. "And the hot oil's very good." The latter would help with a brutal cold triggered by a nonstop cross-country schedule of analysts' presentations he had just concluded.

Whether it was his company or someone else, Lapidus was adamant that by 2010, the $1,000 genome would be within reach—at least wholesale, with maybe a few thousand dollars retail cost to customers. "And it doesn't stop there. It becomes a $100 genome," he said. Lapidus believes there are two fundamental principles in the genomic medicine era. "Each of us has a right to know our genome. No government agency can tell us you can't learn your A's, C's, G's, and T's. The government can and ought to regulate medical action based on that knowledge, but not the knowledge." And second, "You also have an absolute right to privacy. Hacking someone's genome, representing someone's DNA as yours and having it sequenced, and asking for it in any situation with coercion, is not allowed."

"There comes a time very soon when each of us is sequenced once in our lifetimes," said Lapidus, who is PGP-8 in the Personal Genome Project. Many people will pay that $1,000 out of pocket before it is covered by health insurance. That sequence information is essentially fixed, immutable throughout one's life, although the ability to interpret our genetic code will continue to improve for decades.[33] Whether the deliverer of the sequence is the same entity as performs the interpretation is an open question. Lapidus is skeptical whether 23 et al. "can be an expert on all the good and bad things that mankind has that's associated with the genome. They can write a paragraph, they can't write a book chapter [on every disease]." Lapidus spoke from experience: he is a carrier for an inherited form of colon cancer, so knows he is at increased risk of the deadly disease. If he should get the disease, his family would need a full investigation requiring genuine medical expertise. How about a model where sequencing labs provide the basic sequence via a thumb drive, DVD, or Web download, leaving the annotation and interpretation to groups of experts? "Imagine the *Mayo Clinic Guide to the Genome*, or the *American Cancer Society Guide* focusing solely on cancer predisposition," said Lapidus.

There are already promising third-party efforts to interpret genome data, such as Promethease and Trait-o-matic, but Lapidus is intrigued by another model. The *Encyclopaedia Britannica* dominated the market for 200 years or so, with marketing based on expensive binding, gold covers, and an unspoken element of fear. ("Parents, do you want your kid to go to Harvard?") Lapidus recalled his parents bought the cheaper $900 version rather than the deluxe $5,000 version. "Two things happened. One is that it became a CD package, so selling it for $999 was nuts! But one thing *Encyclopaedia Britannica* was, and still is with a capital A, was authoritative. Authorities write the articles. You won't find wacky things in there."

The other blow to *Encyclopaedia Britannica*'s business model was the growth of Wikipedia, a free resource with contributors of varying authority. "Maybe the alternative model is a moderated-for-a-fee wiki, where experts are invited to make submissions." Lapidus continued: "Now pretend you're an MD, counseling an informed patient with a mutation. Who gets tested in my family? Do I get an annual colonoscopy? Do I get my colon removed? It's very easy to make that mistake. The doctor needs to learn more than the patient knows in a very short amount of time." Unfortunately most doctors know little about genetics, and no matter how well informed, a patient can't write a prescription or perform a surgical procedure. But in the world of genomic medicine, there are many things that people can do, including guiding their choice of partner, whether to have a baby or carry it to term, or elect to have an annual colonoscopy.

The information shortage is exacerbated by the current model of genetic counselors employed by consumer-genomics companies, which won't scale and won't work. There simply aren't enough genetic counselors, or "people in the business of helping you worry."

But what are the alternatives? During the early days of telephony, when phones were plentiful but no calls could get through, we saw the invention of the rotary dial. For medicine, the alternative has got to be information over the Web and better-trained physicians.

Lapidus asked me to give the pro and con arguments for personal genomics. I tell him a genome scan provides the most accurate and comprehensive picture available of someone's propensity for rare and common diseases. Every disease is a genetic disease, and the insight obtained is often predictive and actionable; you can make lifestyle and/or medical changes. The converse argument is: We've never met you; you're just a DNA sample. We don't know your environment or family history. Lapidus suggests it is in fact a civil rights argument, a human rights argument: "Do we or don't we have a right to know our As, Cs, Gs, and Ts?"

Until 2010, the FDA showed little interest in regulating personal genomics, which Lapidus attributes to "a mentality of unreasonable caution." His frustration with regulatory agency red tape is long-standing. While Lapidus has been financially successful in his biotech career, his son's father-in-law, a cigarette manufacturer, has apparently done better. One New Year's Eve, he collected Lapidus's family in his private jet and flew them to his home in Oklahoma for a party. Lapidus found it ironic that someone "in the cancer business" could have his own jet. "I've been in the cancer business: pap smears, colorectal cancer. In each of those businesses, the government has thrown innumerable obstacles in front of me. [We] spend 80 percent of our time and money on regulatory and reimbursement issues—issues that my cigarette-making in-law doesn't need to. The irony is that the government puts no obstacles in front of you if you're in the cancer business and puts huge impediments if you're on the other side."

We are moving into a different age, a different era. The days where medicine was practiced locally, and the only outsourcing was reading radiology films, have almost gone. Genomics has no local component, no intimate relationship with a physician for the first step. All things being equal, Lapidus says he'd rather have his test run by an appropriately regulated domestic laboratory, but if the FDA should ever clamp down on direct-to-consumer testing, Lapidus said the industry will simply move offshore.

"If consumers think this is valuable, guys in India will open up All India Genetics Testing—you get your A's, C's, G's, and T's on the Web. The government will be able to regulate American labs, but it won't be able to regulate personal genomics."

CHAPTER 13

The Rest of Us

In 2007, the editors of *Nature Genetics* invited scientists from around the world to post answers to a simple yet tantalizing question: What would you do if the $1,000 genome became a reality?[1] "What wouldn't we do?!" enthused Francis Collins, director of the NIH and the former head of the National Human Genome Research Institute (NHGRI). "We'd be like kids in a candy shop—there are so many exciting possibilities."[2]

Not all of those possibilities involve sequencing *Homo sapiens.* Dog lovers would study the contrasting personalities of pit bulls, golden retrievers, and the like, shedding light on the hereditability of human behaviors. "Is it possible that this approach would allow us to understand the molecular basis of forgiveness and commitment—two behavioral traits shared uniquely by dog and man?" asked NHGRI's Elaine Ostrander. Others suggested sequencing the 38 species of cats to understand patterns of genome diversity, the 375 living species of primates to chart the evolutionary back story of *H. sapiens*, and the 100 most endangered mammals, in case future generations succeed in developing a version of the technology espoused by Michael Crichton in *Jurassic Park*. And the prospect of cataloging the evolutionary relationship of species up and down the tree of life by a sort of genetic bar coding, mused

Rockefeller University president and Nobelist Paul Nurse, "might bury the creationists and the intelligent designers under a mountain of base pairs."

Nurse was also interested in the so-called microbiome—the teeming mass of microbial life that lives on and inside us—in the gut, skin, nose, and so on. The medical relevance of one such bacterial hitchhiker, *Helicobacter pylori*, was revealed only when Australian researcher Barry Marshall drank a Petri dish of bacteria to prove their guilt in the pathogenesis of stomach ulcers. A rather more systematic analysis of microbial populations cohabiting with humans will be a crucial complement to genomic studies on human health. Michael Rhodes, an executive with Life Technologies, proposed "the Gaia Genome Project"—sequencing all known species, effectively the earth's genome. Researchers could submit the DNA of their pet organisms to a centralized sequencing facility—everything from bacteria and protozoa to threatened species on the conservation watch list—in a sort of international Noah's ark program.

For our own species, the $1,000 genome conjures up the enthralling prospect of being able to study our biology and history like never before. Sampling the incredible genomic diversity of indigenous populations around the world, particularly across Africa, would paint a true picture of human evolutionary history. "It is finally time to study all the normal variation that enriches the human world and experience—memory, behavior and personality . . . to understand who we are and how we got that way," said Duke University's David Goldstein. As human populations spanned the globe, many DNA segments were conserved, subjected to forces of positive natural selection. Harvard's Pardis Sabeti has found a way to systematically identify these genome regions, pinpointing gene variants that have played a key role in human adaptation, sensory perception, and resistance or susceptibility to infectious disease.[3] On the flip side, analysis of naturally terminated fetuses, which are ethically acceptable as research subjects in India, could reveal genes that are critical for normal embryonic development.

The same idea could be applied at the cellular level by constructing an "ontogenetic tree" of the human body, cataloging what one contributor called "the rare spontaneous mutational needles in the genomic haystack." In addition to the estimated eighty to one hundred new germ line mutations introduced during each generation, any two cells chosen at random in an individual are likely to be distinguished by several dozen mutations, some of which can provoke cells into the cacophonous growth of cancer. In sequencing the genome of a lung tumor, British researchers documented thousands of mutations and concluded that one mutation is fixed for roughly every fifteen cigarettes smoked.[4]

Then there's epigenetics, a still poorly understood phenomenon whereby the genome is studded by chemical modifications (chiefly methylation) that naturally mute the activity of certain genes, also linked with some cancers. With new sequencing technologies, researchers would love to compare the methylation pattern in stem cells versus normal cells, newborns versus centenarians, identical twins with different disease profiles, and a healthy brain versus an Alzheimer's brain. A person's epigenotype could provide clues to his or her lifetime toxin exposure and the development of diseases such as autism, Alzheimer's, and schizophrenia.

The medical applications of the $1,000 genome are almost boundless. For $100 million, Collins suggested sequencing 100,000 people. Take thirty common diseases, including cancers, asthma, heart disease, and mental illness, and sequence 2,500 patients apiece, unearthing the critical predisposing DNA variants. Then sequence 25,000 healthy centenarians to explore the genetics of good health and longevity. With some 4,000 Mendelian genetic traits cataloged (the molecular basis has been found for about 1,500 of these), about one in twenty couples have a high risk of having a child affected by a Mendelian disorder. Will history repeat itself? Just as medicine has rid the world of polio and smallpox and effectively prevented genetic diseases such as thalassemias in parts of the Mediterranean world, some believe the $1,000 genome will crush or even eradicate most kinds of genetic disease. However, one geneticist warned,

"Like a candle in the wind, their legend will probably continue long after they are burned out."

The $1,000 genome means that genome sequencing is destined for the doctor's office as more and more patients have their DNA sequenced as part of their routine medical care. Jonathan Pritchard, from the University of Chicago, said:

> As a parent, would I have my three-year-old son's genome sequenced to help determine his genetic liabilities (and strengths)? Obviously, this raises serious ethical questions, both for the patient, for people who share his genes, and for a society without universal health care coverage. There will also be serious practical challenges of interpreting the genetic data. But it is hard to believe that the clinical value of such information will not ultimately outweigh the risks.

Cancer investigator John Quackenbush looked forward to the day he could set up a genome browser at home and look for gene variants linked to disease, using that information to improve health through lifestyle modifications, vigilant screening, and prophylactic medications. But as more and more of the global wealthy elite choose to be sequenced, Goldstein predicted they would hire "sequence consultants"—genome personal assistants—to advise them about what their complete sequence means for their health "or, in California, their genetically tailored diets." Goldstein hoped that the benefits of genome sequencing would extend beyond those who can afford to hire personal-sequence consultants.

It is easy to assume that patients, particularly in the United States, accustomed, even entitled, to state-of-the-art medicine, will embrace the $1,000 genome. But how much information do patients really want to know? How will clinicians react when patients insist on their genome data, with or without insurance? One school of thought suggests that a complete genome simply represents too much information for practical use in a clinical setting: there are too many ambiguities and too much potential bad news for patients to deal with.

But many patients will undoubtedly demand the maximal information about their disease risks, the safest and most effective drugs for their genotype, and other health-positive lifestyle changes. Physicians will have to communicate such information to their patients in ways that are meaningful and understandable, as well as keep their patients up-to-date as new disease links emerge. The Centers for Disease Control's Muin Khoury argues that the only way to evaluate the effectiveness of personal genomics is to run a clinical trial to see if volunteers who have had their genomes sequenced use that information to improve their health. What is the benefit of early intervention in prostate cancer? Do good results in type 2 diabetes and heart disease lead to a false sense of security? But others counter that clinical trials weren't deemed necessary before launch or FDA approval for mammography, C-reactive protein, or MRI tests.[5] Unfortunately, it would take decades to fully assess the success of such a trial. And it bears repeating that studying the impact of genomic information should not come at the expense of studying lifestyle and environmental factors.

A few of the respondents to the *Nature Genetics* question had more whimsical ideas for how to spend their $1,000. Bath University's Laurence Hurst suggested studying atonal people who cannot distinguish Bach, Beethoven, or the Beatles from white noise. Perhaps the mystery DNA variants responsible are also involved in language or deafness. From Sydney, Peter Little was in favor of spending some taxpayer dollars on subjects that interested the general public, not just scientists and clinicians, such as the genetics of beauty and happiness. And Canadian geneticist Stephen Scherer dreamed of decoding the DNA of Albert Einstein and Ted Williams to gain a glimpse into "the minimal code for a brilliant mind and the indices for the perfect swing." Neither Scherer nor Little would suggest that happiness or the ability to slug a home run could be explained by a few letters in our 3-billion-base genetic encyclopedia. But how much of our destiny is dictated by our DNA?

In 1997, New Zealander Andrew Niccol wrote and directed the science-fiction movie *GATTACA* (the title is spelled out from the four letters of DNA). Niccol's debut film depicts a near-perfect society that created genetically "perfect" embryos, eliminating disease genes, and enhancing behavioral, intellectual, and physical characteristics. By contrast, Ethan Hawke's character, conceived the old-fashioned way, has a list of flaws, including a heart defect and severe myopia. At one point, Uma Thurman secretly takes a strand of his hair to a DNA dispensary, which instantly prints out his full sequence. At a fertility clinic, a young couple deliberates over which of their four embryos to implant, with the understanding that engineering complex traits such as heightened musical or mathematical ability costs extra. "With multi-gene traits, there can be no guarantee," counsels the geneticist. "You'll probably do just as well singing to him in the womb." The dual-caste-society idea was extended by Princeton geneticist Lee Silver in *Remaking Eden*, in which he speculated that by 2350, the human race could divide into the GenRich, genetic aristocrats replete with synthetic genes, and the Naturals. Eventually they would become separate species.[6]

The embryo selection process depicted in *GATTACA* is based on preimplantation genetic diagnosis (PGD), a method developed in London in the 1980s that is becoming increasingly popular.[7] In PGD, a single cell is teased apart from a three-day-old in vitro fertilization (IVF) embryo, the DNA amplified and screened for the presence of an abnormal gene, before the chosen embryos are implanted into the mother. Traditionally PGD is used to identify for implantation embryos that are free of the genetic defect, asymptomatic carriers of a recessive disease, or females in the case of X-linked disorders that affect only males. PGD spares would-be parents the trauma of contemplating abortion if prenatal diagnosis revealed an affected pregnancy, but it also triggers a separate ethical debate over the fate of frozen spare embryos, many of which are ultimately destroyed with parental permission.

On the thirtieth anniversary of the birth of the world's first IVF baby, "SuperBabe" Louise Brown, *Nature* imagined the birth of a beautiful baby girl thirty years hence, one that involved screening by PGD, complete genome sequencing, and selection by her parents "when the Baby's First Four Letters analysis at the clinic said that this particular embryo had the best odds of growing up to be thin, happy and cancer-free."[8] Even with personal-genome sequencing heading toward the fertility clinic, *Nature* dutifully acknowledged that parents couldn't actually design a baby or guarantee their embryos a perfect genetic future. But the techniques could grant parents the opportunity to create a top-five wish list of desirable characteristics for their child and select the embryo that conformed most closely to those attributes. (Confusingly, *Nature* went on to say that the wish list might include such attributes as intelligence and ambition, rather contradicting itself.)

In his book *Genetic Destinies*, Peter Little offered his own vision of a baby girl born around the year 2020.[9] After counseling in advance as to the range of their daughter's cosmetic, personality, and intellectual features, the child's parents conceive ten test-tube embryos by IVF. Embryo #10 looked very promising: red-haired and freckled, an extrovert personality, with tolerable risks of cancer, Alzheimer's disease, depression, and schizophrenia and, as a bonus, HIV resistant. After birth, the baby is vaccinated against common cancers and diseases based on her genotype. She is also tested for scores of genes on the GlaxoSmithMerckNovartis drug-response list so that future drug treatments match her genes. This information is encoded on a microchip and implanted under her skin. She proceeds to live a healthy life, with occasional injuries treated by stem cell transplants and gene therapy, to the ripe old age of 120.

In the near future, couples using PGD or sperm banks could be presented with a set of fully sequenced embryos with various permutations to choose from—free of this disease, a greater chance of that physical trait. Of course, one day parents might have to explain to their children why they made the choices they did.[10] Not that it is likely that PGD will supersede sex as the preferred means to have babies—IVF is expensive and much less pleasurable—but its use for more and more disorders is growing fast and threatening to spill over into nonmedical conditions. Moreover, it is technically feasible to extend this screening to hundreds or thousands of genes from a single embryo—or even the full genome.

At one of the leading PGD clinics in the world, the Genesis Genetics Institute in Detroit, medical geneticist Mark Hughes has screened for 250 different genetic disorders and adds one or two every month, some on request.[11] "We can pretty much do anyone as long as the molecular basis is known," he said.[12] He routinely tests for Ashkenazi Jewish disorders (Tay-Sachs, Bloom syndrome, and Canavan disease, for example); classic genetic diseases such as cystic fibrosis, Huntington disease, and sickle-cell anemia; rare disorders such as Hallervorden-Spatz disease and Simpson-Golabi-Behmel syndrome; and several cancers, including breast cancer (*BRCA1, BRCA2*).

Hughes adroitly navigates the ethical minefields—for example, selecting embryos lacking the *BRCA1* mutation reduces, but does not eliminate, the baby's risk of eventually developing breast cancer.[13] Genesis also tests for Alzheimer's disease, but a rare familial early-onset form of the disease, not the common *APOE4* variant. And Hughes helped pioneer the use of PGD to screen embryos who could provide life-saving stem cells from umbilical cord blood for an affected brother or sister—what the British tabloids call "savior siblings." "I've designed my baby," was how one UK tabloid headline writer described a mother's decision to use PGD to spare her child from inheriting the deadly eye cancer retinoblastoma.[14] Hughes staunchly defends this practice: one of his patients was the wife of basketball star Carlos Boozer, who gave birth to healthy twins selected by PGD and provided a cord blood transplant for their oldest child, who had sickle-cell anemia.

Hughes says any fertility clinic straying into the area of screening for nonmedical traits such as skin or eye color would immediately be ostracized, but at least one clinic in Los Angeles claims to be heading down that road, and a London clinic conducted PGD for congenital fibrosis of the extraocular muscles—in lay terms, to spare the newborn from being severely cross-eyed.[15] Hughes doesn't think PGD should be used for trait selection, but believes there's a big gray zone between a trait and a disorder, and squinting probably falls in that zone. Said Hughes, "With $1,000 genomes, now people are going to be aware of all the genes they carry, and they might be looking for technologies other than throwing genetic dice to have their kid."

Another PGD ethical minefield is deliberately selecting embryos with a defective gene, such as for dwarfism or deafness.[16] It sounds unthinkable, but a few PGD clinics have reportedly been asked to consider "negative enhancement"—selection of an embryo so it has the same disease or disability as the parents.[17] Also controversial is sex selection, which some IVF clinics offer under the guise of family balancing—something Hughes condemns as a mere marketing ploy. "That's not medicine," he said. "I'm not sitting on my high horse telling people it's evil, but I'm too busy taking care of people who are sick, and I'm not doing that sort of thing. There's no pathology, there's no pain and suffering. I don't see any reason why doctors should be involved in it."

But others justify the ability to select for complex traits such as clear skin under the guise of cosmetic medicine. Clearly as technology improves, the capacity to screen for multiple genes will exist. Hughes said it's already possible to analyze up to 150 genes per embryo. "We don't need to, and on the horizon I don't see any reason why anyone would do it. But you certainly could," he said. For that to work, one would need dozens of IVF embryos to compare and select the desired combination of genes. That's not practical today, but Hughes points to parallel advances in fertility treatments such as in vitro maturation, which allows women to harvest many eggs for safe freezing while they're young so they can be matured in the lab a decade or two later.

"One man's optimizing outcomes is another man's eugenics," suggested Helicos founder Stan Lapidus.[18] "I want a beautiful child, not bald, no love handles, no glasses, smooth skin—presumably it will be possible to partially select those traits. Is that part of our future landscape? It's not a stretch to think that educated people, who currently use computer dating services like eHarmony and Match.com, could eventually use genomics to refine that search."

Lapidus recounted the story of his friend Wyc Grousbeck, a Princeton graduate and champion rower, venture capitalist, and co-owner of the Boston Celtics. The "movie star handsome" Grousbeck married "a stunningly beautiful woman," and everything turned out the way it should, he said—except their son is blind. The Grousbecks both carry a recessive gene for Leber's congenital amaurosis, a rare hereditary form of blindness. They relocated to Boston to enroll their son in the world famous Perkins School for the Blind. Would one choose to have a disabled child and grow through the experience, Lapidus wondered? What if Grousbeck had never met his wife, or married some other beautiful, well-educated woman? Or they'd decided to have PGD to screen for mutations? he said. "Imagine a version of Facebook or eHarmony or JDate that you voluntarily submit your genome. . . . 'Look, here's a bunch of recessive diseases that I surely don't want to be matched up with. If I'm a CF carrier, don't pair me with a CF carrier.' "

There are many other advances in genetic screening as the cost of genome analysis falls. Born on Valentine's Day 1997, Hunter Kelly was diagnosed with an incurable inherited disorder called Krabbe disease, one of a class of genetic diseases known as the leukodystrophies that includes the disease portrayed in the film *Lorenzo's Oil*. Given little chance of reaching his second birthday, Hunter bravely battled for eight years before his death. His father, Jim Kelly, the former Buffalo Bills quarterback, launched the Hunter's Hope Foundation to campaign for the adoption of consistent, comprehensive newborn screening across the United States, which could save thousands of lives each year. "We may have never won a Super Bowl, but my Super Bowl win will be when every child in every state is screened for every treatable disease. It's the losses that we learn from—and no loss compares with losing a child," said Kelly.[19]

Newborn screening was pioneered in the 1940s by Robert

Guthrie, who, having developed a simple blood test for a form of mental retardation called phenylketonuria (PKU), proposed the newborn heel stick test to test for PKU and other "inborn errors of metabolism." Today the simple blood prick is the basis for genetic testing of more than 90 percent of American newborns. Of the more than 4 million babies born annually in the United States, about 10,000 will test positive for one of the conditions on the newborn screening panel. But that testing is currently a mosaic patchwork that makes screening for a potentially fatal genetic disorder something of a zip code lottery. For example, two states, Massachusetts and Washington, do not mandate testing for a potentially life-threatening disorder, 3MCC deficiency, a failure to metabolize the amino acid leucine.[20] Mississippi currently tops the state charts by screening for fifty-seven diseases, while Texas and Arizona screen fewer than thirty disorders.[21] The March of Dimes estimates that each year, more than a half-million babies are not screened for the full panel of life-threatening genetic disorders, costing the health care system more than $1 billion a year—the cost of managing serious and/or chronic complications such as mental retardation.

The Guthrie heel stick test currently relies on mass spectrometry to detect various key metabolites and proteins in the baby's blood sample. But the day is fast approaching when the same non-invasive method could be used as the basis for a comprehensive genome scan, complementing or even replacing the current test. This could take place at birth or be offered to couples before conception. A California genetic testing company, Counsyl, is striking a blow against preventable genetic disease by offering a test to detect carriers of more than 100 genetic disorders for $349. The test includes more than 100 different cystic fibrosis mutations and the full Ashkenazi panel of disorders. Counsyl advisors include Harvard's Steve Pinker and Henry Louis Gates.

An even more extensive screening panel is the goal of the Beyond Batten Disease Foundation, established by Texas biotech entrepreneur Craig Benson and his wife Charlotte. A few years ago, the Bensons were distressed to learn that their daughter had Batten disease, a rare, inherited, fatal disorder of the nervous system. "The day we got Christiane's diagnosis was a horrible, obviously very painful day," said Charlotte. "Disbelief, despair, sorrow, agony, fear, and unbearable pain. It forever changed our path." Benson's foundation aims to eradicate not just Batten's but as many as 500 of the most severe, fatal inherited disorders—the long tail of genetic disease—by sequencing all the relevant disease genes—about 2 million bases in total. The Beyond Batten test would be aimed at couples considering pregnancy, similar to the successful screening programs in the Ashkenazi Jewish community, or adolescent girls following their first OB/GYN appointment.[22]

Just like John Crowley, the hero of the film *Extraordinary Measures* who developed a treatment for Pompe disease, the Bensons, the Kellys, and the Grousbecks had no way of knowing both partners carried a rare genetic mutation—in the same gene. Such tragedies have been sharply reduced in certain populations for ethnic or geographic reasons and increasingly can be avoided. We vet our potential partners by their physical manifestations, so why not screen by DNA as well? Although it might decrease the pool of eligible candidates, Lapidus said, "It would increase the probability of a biologically successful, if not romantically successful, union immeasurably."

In addition to the most life-threatening genetic disorders, screening in the $1,000 genome age would identify patients with more ambiguous, harder-to-diagnose diseases. In 1996, Dennis Drayna's team at Mercator Genetics identified the gene affected in hemochromatosis, one of the most common recessive disorders in people of European descent. The iron overload condition is treated the same way today as centuries ago: regular phlebotomy, or bloodletting.[23] The cost of the test is about $100, but according to Drayna, "The powers that be, in their infinite wisdom, have chosen against population screening for this mutation, even though I think you're just absolutely out of your mind not to screen the population."[24] The counterargument is that genetic screening will ruin health insurance coverage, potentially exposing patients and carriers to "labeling" and insurance repercussions. But Drayna, who is now at the NIH, says he is tired of reading obituaries of people dying needlessly of secondary liver failure or a stroke because they have suffered irreversible organ damage. "Why these people aren't being diagnosed earlier is a travesty," he said. He once asked a prominent geneticist how she felt about the widows of hemochromatosis victims, and she replied it would be wrong to expose hundreds of millions of people to genetic discrimination. "Puh-leeze," said Drayna. "We're equating the potential problems with discrimination to death and leaving teenage children in the lurch? Give me a break."

The idea of universal screening—"newborn screening for adults" as he calls it—was a huge factor that sparked Dietrich Stephan's decision to launch Navigenics. Stephan wants to be able to sequence an individual's genome at birth and unmask various results, beginning with the newborn screening panel, as the child reaches various milestones, such as turning eighteen or twenty-one. That information will allow a physician to suggest medical or lifestyle decisions and take other screening measures. Interestingly, his Navigenics cofounder, David Agus, doesn't quite see it that way, citing concerns over the consent process for newborns. Some patients want to know everything, others nothing. "It's your right. I'm not here to say everybody should have genetic testing. Parents shouldn't make that decision now for children. It's too early." But something like the Navigenics software could provide the medical interpretation for the genome data produced by the sequencing companies, possibly piped directly to physicians.

While there are certain philosophical differences between 23andMe and Navigenics, Anne Wojcicki cannot hide her enthusiasm for the benefits of universal newborn screening, even as she acknowledges the plethora of ethical issues that would be unleashed. "We don't want parents saying, 'I just want this super-athletic child.' But we have a number of friends of the company who are severely dyslexic and say, 'It would've been really great as a child to have not felt dumb my first twelve years.' How great would it be if you could say, 'You have a higher chance of dyslexia; you should just watch these symptoms?" When Wojcicki was expecting her first child, her physician told her a placental embolism carries a 90 percent chance of death. "It's really petrifying—if you had that and survived, you don't know if you should have a second child or not."[25]

Some consumer-genomics companies are taking advantage of this sort of parental pride—and fear. "What if you're told that your child could be the next EINSTEIN, BILL GATES, or the next TIGER WOODS" screamed a marketing e-mail I received from My Gene Profile in 2010, apparently undeterred by news of Tiger's adult transgressions. For anywhere from $1,200 to $4,500, the Connecticut firm says it tests for forty traits so that parents can unleash their child's full potential "gifted by God." Among the genes tested are "the intelligence gene," "the sensitivity to second-hand smoke gene," "the faithfulness/loyalty gene," and "the propensity for teenage romance gene." There is no description or justification of those glib descriptions provided on the firm's Web site. I could not even find out what the specific genes were. This places the average consumer in jeopardy of being misled into spending a lot of money on a test that will not make one iota of difference in their progeny's God-given ability to perform calculus, write software, or win the Masters.

Would parents really contemplate loading the genetic dice in choosing their unborn children? George Church believes that in the next few decades, our ability to diagnose disease will outstrip the ability to intervene. He points out that parents think nothing of hiring coaches for their kids or medicating children with mild attention-deficit disorder; which is one definition of designer babies. Parents pay thousands on orthodontics to optimize their children's physical attractiveness, so why not maximize their children's genetic checklist? During one public debate, Eric Lander posited a futuristic scenario whereby couples who increase the prevalence of genetic diseases that burden society should carry more of the financial burden. Imagine telling prospective parents who know their baby will carry deleterious genes: "You're free to choose to have this carrier child, but if you choose to have that choice, we're not going to pay for it. We will pay for IVF."[26]

Although it seems logical to want to eliminate common mutations that predispose to diabetes, obesity, or Alzheimer's disease, the purging of rare deleterious genes—for bipolar disorder, for example—could also result in society's losing some of its most creative individuals, or "compressing the bell curve," as Church calls it. Indeed, mutations sometimes get a bad rap. In some respects, they are good—the fuel for evolution. Just as the sickle-cell mutation gives carriers an advantage against malaria, the prevalent cystic fibrosis mutations might have given carriers an advantage in warding off infectious diseases such as cholera centuries ago. Given the well-established link between manic depression and creativity, Church says parents might face the dilemma of implanting an embryo who might end up being a musical genius or suicidal.

In short, we don't know what might be learned about certain mutations in the future. In the film *Sleeper*, Woody Allen's character, Miles Monroe, the owner of a health food store, awakens from 200 years of accidental cryopreservation requesting wheat germ and organic honey. "You mean there was no deep fat? No steak or cream pies or hot fudge?" one doctor asks another. The other replies: "Those were thought to be unhealthy, precisely the opposite of what we now know to be true!"

A decade after the first draft of the human genome, the cost of sequencing an individual personal genome has plummeted from $1 billion in 2000 to $1 million in 2007 to just about $1,000 in 2010. Those latter figures focus on reagent costs and do not fully include overheads such as power and personnel, data analysis and depreciation, but the trend is clear. And it will not stop there. The next wave of third-generation sequencing technologies looks destined to drive costs down to $100 or less. Increasingly the expense won't be sequencing the DNA but making sense of it. Unless we can bridge the gap between sequence data generation and analysis, then, as Canadian cancer researcher John McPherson pointed out, "the coveted $1,000 genome will come with a $20,000 analysis price tag."[27] However, we are already seeing smart software programmers develop programs to make sense of the surge of genomic data and the rapidly growing knowledge of network and systems biology, helping assign physiological relevance to the variants uncovered by genome-wide association studies (GWAS) and complete genomes.

In the coming years, we will see revelatory discoveries about the human genome, although I doubt that will extend to finding a gene that imparts the ability to hop through time, like Eric Bana in *The Time Traveler's Wife*, or one that determines our religious faith, which was the extraordinary impression conveyed by Dean Hamer in *The God Gene*.[28] But scientists are going to continue making dramatic advances in our understanding of the book of life. I'm writing this the same month that scientists reported a weak spot in the genome linked with obesity and Dennis Drayna's group identified gene mutations associated with stuttering.[29] More and more discoveries about the role of our genomes and the environment in our health and disease susceptibilities will keep tumbling out, along with breakthroughs in crunching and integrating these data to build computational models that provide a picture of the true inner workings of our cells and tissues.

I started this chapter discussing just a few of the myriad applications of the $1,000 genome in the research realm, but opinions remain divided over whether $1,000 for a complete genome sequence—which, after all, is about the same price as two or three routine genetic tests today—is worth the investment. Throughout this book, I have noted the many reasons for and against personal genomic testing. Cost will remain an issue for many but those objections will fade as the cost inexorably drops below $1,000. Reimbursement by health insurance companies is another big issue, but there are signs that more genetic tests are being covered. (The cost of screening is trivial compared to the toll of managing a life-threatening genetic disorder or a chronic late-onset disease.) Some may be worried about news they might learn, a concern loudly voiced by the medical establishment, which argues that genomic self-empowerment will mislead consumers and could cause harm—from anxiety about a test result to a false sense of security to unnecessary surgery—although such fears remain entirely hypothetical at this point. And there is no good reason that personal genomes should be held to a higher standard than the drug industry. If medical safety is the issue, then the FDA would never approve a single drug, for as one leading pharma executive said a few years back, "There is no such thing as a safe drug."[30] The Genetic Information Nondiscrimination Act has allayed most concerns about the privacy of genetic information in the United States, and no one is suggesting that personal-genome testing should be compulsory.

Perhaps the most important issue for treading carefully is our still nascent ability to interpret our personal genetic code, which is complicated and probabilistic. To be fair, some pretty smart people think it's a waste of time. Nobel laureate Sydney Brenner was one of the first people to gaze in awe at Crick and Watson's iconic model of the double helix in 1953, but he is highly skeptical of personal genomics and probabilistic medicine. "Would I want to know? Would I want to know *what*?" he mused, before telling a story.[31] Astronomy began centuries ago as a pure observational science mapping the heavens but forked into two paths: the first gave us astrophysics and cosmology, the other astrology. The stars can determine your future, telling you that when Jupiter is in conjunction with Venus, you will meet a dark stranger. By analogy, Brenner continued, "it's not genomics but genonomy. We have genetics, which is the science . . . and we've generated another science, which I call 'genology.' That's the science that says, 'If you've got this SNP and that SNP, you are going to become a dark stranger!' "

Brenner takes a dim view of people who might enjoy going to cocktail parties and saying, "Oh, by the way, I had my genome done the other day. Isn't it nice?" Nor does he see the point of sharing information or comparing sequences on the computer like an iTunes library. Genome variations "purport to predict what will happen in the future," but except for rare cases such as Huntington's disease, "it's not the medicine of certainty." Of course, Brenner is all for genome research to unlock the gold mine of information buried within our species and those of our evolutionary cousins.

That said, almost all medicine is probabilistic. Measuring blood pressure, cholesterol, or prostate-specific antigen does not guarantee any particular health outcome. Few people can identify with former NBC correspondent Charles Sabine, who was diagnosed with Huntington's disease. His brother, who also has the incurable disease, practices law in London but now has to practice walking in a straight line before he visits his neurologist. Sabine said he was "absolutely terrified" when he learned his diagnosis, scared that he wouldn't be able to dance with his daughter on her sixteenth birthday. But he said, "Knowing my genetic condition both empowers and motivates me to be part of that battle" to conquer the disease. "Nothing incentivizes more than knowing your genetic code."[32]

There is another reason *not* to rush into personal genomics—one that was eloquently voiced by Microsoft founder Bill Gates. If personal genomics is defined narrowly as rich people having their genes sequenced, Gates argued that wouldn't provide much benefit to the Third World. "Any time you divert $1,000 away from spending on help for the poor, you basically lose a life. We can save lives for $1,000 by improving the vaccination system in developing countries."[33] Those sentiments were echoed by Sir John Sulston, who led the UK's arm of the Human Genome Project. "I don't want my genome sequenced, not because I don't care . . . but I know it will be wasting resources that ought to be spent on global health, not on feeding egos of rich westerners."[34] But if personal genomics is defined more broadly as making genetic information almost free for scientists to gather, then "the benefit of this revolution for health conditions in the poor countries will be very, very dramatic." Gates said the biggest unsolved challenge in global health is the deep inequity that sees young people in poor countries fall victim to infectious diseases. "Genetic understanding will let us create cures to help us get rid of the top 20 infectious diseases." And once that happened, Gates pledged he would finally get his own genome sequenced.

That puts Gates at the back of the queue—a long way behind personal genomics trailblazer Anne Wojcicki, whom I caught up with shortly after her son Benji's first birthday.[35] "It's been a lot of work but a lot of fun," she said of trying to juggle motherhood and management. I strongly suspected Benji was already a 23andMe client, but had to ask. "Of course we tested him!" Wojcicki said, though she declined to divulge any of his results, including his potential risk of Parkinson's. "I think because he's a baby and he's a minor, we respect his own privacy," she said. 23andMe was biobanking the leftover DNA of its clients' saliva in preparation for what Wojcicki called the inevitable offering of a sequencing product. Would her first grandchild have a full sequence at birth? "Yes," she replied unhesitatingly. Perhaps I should have asked about Benji's future brother or sister instead.

Indeed, the Brin grandchild will be one of a new generation who browse (Google?) their genome as avidly as their Facebook page (or whatever social networking outlet then is cool). With 6 billion genomes on the planet and 60 million new ones born every year, not to mention the tumors that in essence are their own genomes, the demand and the need for genome sequencing is destined to explode. No wonder there are so many talented scientists vying to build the technology that decodes all that DNA. The cost won't be $1,000; it will be $100, or $50. Or less. "I want to build a $10 sequencer that can sequence the genome, and do it for maybe $10," said Stanford professor Ron Davis. Based on the technology I've reported on and the people I've interviewed for this book, I am convinced this is coming.

At the end of April 2010, just a few days after National DNA Day and approaching the tenth anniversary of the first draft of the Human Genome Project, a remarkable gathering of personal-genome pioneers assembled for a special one-day conference.[36] Alas, there was no sign of Craig Venter or Archbishop Desmond Tutu, nor was there a surprise appearance from Glenn Close or Marjolein

Kriek. Thus, George Church's audacious plan to gather, for the first and only time, the charter members of the human genome club—the pioneers who have had their genomes fully sequenced—did not entirely come off. But nearly everyone else was there.

There were Jim Watson and fellow scientists Jim Lupski, Stephen Quake, and Seong-Jong Kim from Korea; CEOs Jay Flatley (Illumina), Greg Lucier (Life Technologies), and John West and his eleventh-grade daughter Anne; Personal Genome Project volunteers Misha Angrist, Rosalynn Gill, Esther Dyson, and Kirk Maxey; venture capitalist Hermann Hauser; and Harvard professor and filmmaker Henry Louis Gates.

Watson, outspoken and impatient as ever, wanted to see 100,000 people sequenced as soon as possible, even though interpreting the data would be expensive. "It'll be like motherhood—just do it!" he said. When not showing off his genome on his new iPad, Jay Flatley said that wide-scale whole-genome sequencing was inevitable. "We talk about privacy, ethics, bad things. We need to get far enough along where we tip the balance, where the advantages [of sequencing] are so overpowering the risks of sequencing become minor in comparison." Greg Lucier, his counterpart at Life Technologies, was also eyeing the future role of the genome physician, which he expected to become a vast new subspecialty, like radiology.

"We're ten years away from having every baby sequenced and [the data] becoming part of the electronic medical record," said Esther Dyson. "The utility is going to become unquestioned in the next few years." She took a swipe at the state and other authorities who want to put barriers between people gaining access to their own genetic data. "It's like you can't talk to God yourself, you need a priest," she said. As for the concern that some people might have difficulty understanding the probabilistic nature of the information, she had a simple retort: "If people can understand stats about the Red Sox, they can understand about genetics." Within a couple of years, Dyson all but guaranteed that 23andMe would be offering a personal sequence service. "We won't be doing SNPs. Why bother?" she said.

When Dyson, who had been training to be a cosmonaut, was asked about the risks of genome analysis, former Solexa CEO John West had a rejoinder: "Esther wants to get shot into space on a Russian rocket and you're asking about the risks of her genome?!" he said.

Henry Louis Gates, the African-American host of the PBS series *Faces of America,* talked about "exploding the myth" of racial purity. His own grandfather was so white "we used to call him Casper!" joked Gates. One of the highlights of having his and his father's genomes sequenced was that it allowed him to see the portion of his genome that was inherited from his late mother. "It was like my mother coming back from the grave! These data can have an emotional contact," just like a photograph.

Anne West, John's remarkably self-assured daughter, shared her experience analyzing her own genome, a dedicated endeavor considering the only tool she had to work with was a Microsoft spreadsheet. Her father paid for his entire family to be sequenced using the Illumina personal-genome service, close to $200,000, an idea triggered in part by an embolism he had suffered a few years earlier.

As the genome pioneers shared their experiences and fears, hopes and dreams, I noticed a couple of small children running around just outside the conference room, amusing themselves with video games and plenty of free food. They turned out to be Subio (age five) and Jaesu (age three) Bhak, the children of Jong Bhak, the director of the Personal Genomics Institute in Korea, and his American (Caucasian) wife. They were also the youngest members of the genome sequencing club. "We wanted to know how different two siblings can be," said Bhak.[37] "Our hybrid kids can give us some easy confirmation on that. They are brothers, but their genetic makeup will be much more different from each other than any two random people in the same population."

Bhak said there was a social reason for sequencing his sons as well. "They are very young and do not know what DNA is exactly. However, my guess is that in ten years' time, humans will have their newborns sequenced for various reasons." He wanted to pioneer the social attitudes of growing up with a fully known genome sequence. "We can tell the world the pros and cons ahead of time," he said, adding that he hoped his two "genome kids" would appreciate his actions and not feel upset.

It was one more sign that we are about to enter a new era in medicine and society, when our genomes become as much a part of us as our height, weight, and blood pressure. As Jonathan Rothberg said just before he presented Jim Watson with his digital genome back in 2007, "I think Jim is the first genome for the rest of us." How right he was.

Notes

INTRODUCTION

1. Jeffrey Gulcher, YouTube video, December 5, 2007, http://www.youtube.com/watch?v=pmOyWpAw2vw&feature=channel.
2. Anita Hamilton, "The retail DNA test," *Time,* October 29, 2008, http://www.time.com/time/specials/packages/article/0.28804.1852747_1854493_1854113.00.html.
3. The G2109S mutation is an alteration in which the normal glycine (G) amino acid at position 2109 of the LRRK2 protein is substituted by a serine residue (S). LRRK2 is one of a handful of proteins known to be mutated in rare familial cases of Parkinson's disease.
4. Sergey Brin, "LRRK2," TOO, September 18, 2008, http://too.blogspot.com/2008/09/lrrk2.html.
5. Tip of the cap to Michael Cariaso for coining the phrase "23 et al." In addition to 23andMe, deCODE, and Navigenics, Pathway Genomics launched in 2009. Knome offered whole-genome sequencing for wealthy individuals.
6. David J. Hunter, Muin J. Khoury, and Jeffrey M. Drazen, "Letting the genome out of the bottle—Will we get our wish?" *New England Journal of Medicine* 358 (2008), 105–107.
7. Robert Green, phone interview with author, October 24, 2008.
8. Matt Ridley, *Genome: The Autobiography of a Species in 23 Chapters* (New York: HarperCollins, 2000).
9. Dietrich Stephan, interview with author, March 26, 2008.
10. J. Craig Venter, "A DNA-driven world," The 32nd Richard Dimbleby Lecture, BBC, December 4, 2007.
11. Kari Stefansson, phone interview with author, June 25, 2008.
12. Dietrich Stephan, panel discussion at Advances in Genome Biology and Technology conference, Marco Island, February 7, 2008.
13. Kevin Davies, *Cracking the Genome* (New York: Free Press, 2001).
14. In fact, the cystic fibrosis gene wasn't on chromosome 13. By 1985,

several groups including ours had cornered the CF gene on chromosome 7. Four years later, Francis Collins teamed with a pair of Canadian investigators to identify the gene, to great international acclaim. See Kevin Davies, "The search for the cystic fibrosis gene," *New Scientist,* October 21, 1989, http://www.newscientist.com/article/mg12416873.900-the-search-for-the-cystic-fibrosis-gene-for-nearly-a-decade-several-teams-of-molecular-biologists-have-struggled-to-be-the-first-to-find-the-defective-gene-that-is-responsible-for-cystic-fibrosis-at-last-a-collaborative-effort-has-succeeded.html?full=true.

CHAPTER 1. JIM AND CRAIG'S EXCELLENT ADVENTURE

1. "Nobel laureate James Watson receives personal genome in ceremony at Baylor College of Medicine," Baylor College of Medicine press conference, Houston, Texas, May 31, 2007, http://www.bcm.edu/news/packages/watson_genome.cfm.
2. T. R. Reid, *The Chip: How Two Americans Invented the Microchip and Launched a Revolution* (New York: Random House, 2001).
3. Andrew Pollack, "Company says it mapped genes of virus in one day," *New York Times,* August 22, 2003, http://www.nytimes.com/2003/08/22/business/company-says-it-mapped-genes-of-virus-in-one-day.html.
4. 454's first sequencer lacked the accuracy to tackle a project as daunting as a human genome. Improvements in the second version of the instrument, the FLX, made an attempt feasible in early 2007. The method involved fragmenting Watson's DNA into millions of tiny fragments and then using a computer to reassemble the finished sequence, like a giant jigsaw puzzle. As is customary, 454 scientists sequenced Watson's DNA many times over to ensure that the vast majority of his genome was sequenced at least once. 454 achieved eightfold coverage, meaning it produced 24 billion bases of total sequence compared with the 3 billion bases of Watson's genome. Nowadays it is conventional to strive for at least thirtyfold coverage.
5. Meredith Wadman, "Watson's genome sequenced at high speed," *Nature* 452 (2008), 788.
6. Renato Dulbecco, "A turning point in cancer research: sequencing the human genome," *Science* 231 (1986), 1055–1056.
7. Watson later vented his frustration with the pace and direction of cancer research in an Op-Ed article: James D. Watson, "To fight cancer, know the enemy," *New York Times,* August 5, 2009, http://www.nytimes.com/2009/08/06/opinion/06watson.html?_r=1&sq=james%20watson%20cancer&st=cse&adxnnl=1&scp=1&adxnnlx=1267822919-4tVJV1TAPH4MikLO3dCI1w.

8. The *APOE* gene exists in three "flavors," or *isoforms*—E2, E3, or E4—depending on variants at two positions within the gene. The E4 variant is rare but has been associated with increased risk of atherosclerosis and Alzheimer's disease.
9. Jim Watson's genome sequence was made available online via a genome browser developed by Lincoln Stein at Cold Spring Harbor Laboratory, http://jimwatsonsequence.cshl.edu/cgi-perl/gbrowse/jwsequence/.
10. Dale R. Nyholt et al., "On Jim Watson's *APOE* status: genetic information is hard to hide," *European Journal of Human Genetics* 17 (2009), 147–149. The authors contacted Watson about the possibility of inferring his Alzheimer's risk based on sequence flanking the *APOE* gene; Cold Spring Harbor Laboratory staff subsequently scrubbed some 2 million bases from the Watson genome browser.
11. My first book chronicled the race for the breast cancer gene: Kevin Davies and Michael White, *Breakthrough: The Race for the Breast Cancer Gene* (New York: John Wiley & Sons, 1996).
12. These databases include the Human Genome Mutation Database (http://www.hgmd.cf.ac.uk/ac/index.php) and Online Mendelian Inheritance in Man (http://www.ncbi.nlm.nih.gov/omim).
13. James Lupski, phone interview with author, December 6, 2007.
14. Michele Clamp et al., "Distinguishing protein-coding and noncoding genes in the human genome," *Proceedings of the National Academy of Science* 104 (2007), 19428–33.
15. David Wheeler, interview with author, Providence, R.I., October 18, 2007.
16. David A. Wheeler et al., "The complete genome sequence of an individual by massively parallel DNA sequencing," *Nature* 452 (2008), 872–876.
17. Maynard Olson, "Dr. Watson's base pairs," *Nature* 452 (2008), 819–820.
18. Wadman, "Watson's genome sequenced at high speed."
19. Samuel Levy et al., "The diploid genome sequence of an individual human," *PLoS Biology* 5, (2007), e254, doi:10.1371/journal.pbio.0050254. The significance of the "diploid" genome is that Levy and colleagues obtained sequence from both chromosomes in each pair, rather than settling for a composite "haploid" genome. Put another way, they sequenced 6 billion bases (46 chromosomes) rather than just the 3 billion in a set of 23 chromosomes.
20. Ironically, Venter's 2001 *Science* paper devoted several column inches describing the precautions Celera took to ensure the anonymity of Celera's five DNA donors, which Venter later breached.
21. Kevin Davies, "John Craig Venter Unvarnished," *Bio-IT World*, November 2002, http://www.bio-itworld.com/archive/111202/horizons_venter.html.
22. Antonio Regaldo, "Entrepreneur puts himself up for study in genetic

'tell-all,' " *Wall Street Journal,* October 18, 2006: A1, http://online.wsj.com/public/article_print/SB116113107659595911-B8g1u21x_BL74Yp8fOIIdnBa2RO_20061025.html. Venter also published vignettes about some of his personal gene variants in his autobiography: J. Craig Venter, *A Life Decoded* (New York: Viking, 2007).

23. 454's Michael Egholm presented preliminary details of Project Jim at the Next-Generation Sequencing Applications and Case Studies conferences (Cambridge Healthtech Institute), San Diego, March 21, 2007.
24. Levy et al., "The diploid genome sequence." The team catalogued 4.1 million DNA variants spanning 12.3 million bases. There were 3.2 million novel SNPs, 900,000 indels and CNVs (ranging from 1 base to 80,000 bases in length). Forty-four percent of Venter's genes contained a variant in one copy or another, and 70 percent of the total variation was *not* in SNPs, illustrating that most human genetic variation stems from CNVs and confounding conventional wisdom that we all possess the exact same set of genes.
25. Pauline C. Ng et al., "Individual genomes instead of race for personalized medicine," *Clinical Pharmacology and Therapeutics* 84 (2008), 306–309.
26. James Watson, "Living with my personal genome," *Personalized Medicine* 6 (2009), 607.
27. J. Craig Venter, "The DNA-driven world," 32nd Annual Richard Dimbleby Lecture, BBC, December 4, 2007, http://www.bbc.co.uk/pressoffice/pressreleases/stories/2007/12_december/05/dimbleby.shtm.
28. Erika Check, "Celebrity genomes alarm researchers," *Nature* 447 (2007), 358–359.
29. Olson, "Dr. Watson's base pairs."

CHAPTER 2. 23 AND YOU

1. Gideon Rachman, "Some activities shouldn't involve spitting unless things go badly wrong," *Financial Times,* January 24, 2008, http://www.ft.com/cms/s/0/166f6930-ca1f-11dc-b5dc-000077b07658.html?nclick_check=1.
2. Felix Lowe, "Google wife targets world DNA domination," *Daily Telegraph,* January 25, 2008, http://www.telegraph.co.uk/finance/newsbysector/mediatechnologyandtelecoms/2783261/Davos-2008-Google-wife-targets-world-DNA-domination.html.
3. Anne Wojcicki, interview with author, Mountain View, Calif., June 9, 2008.
4. Amy Harmon, "My genome, myself: Seeking clues in DNA," *New York Times,* November 17, 2007, http://www.nytimes.com/2007/11/17/us/17dna.html?pagewanted=1; and Thomas Goetz, "23andMe will de-

code your DNA for $1,000. Welcome to the age of genomics." *Wire,* November 2007, http://www.wired.com/medtech/genetics/magazine/15-12/ff_genomics?currentPage=4.

5. Drama 2.0, "The dumbest start-ups of 2007," December 23, 2007, http://mashable.com/2007/12/23/the-dumbest-startups-of-2007/.
6. Sarah Boseley, "DNA test website raises accuracy fears," *The Guardian,* January 22, 2008, http://www.guardian.co.uk/science/2008/jan/22/genetics.health.
7. Martin Bashir, "Individual genetic risk factors delivered via Internet," ABC News, February 13, 2008, http://abcnews.go.com/Health/story?id=4284602&page=1.
8. Linda Avey, interview with author, Boston, November 19, 2009.
9. The *Today* show, NBC, March 9, 2008, http://www.thegeneticgenealogist.com/2008/03/15/genetic-testing-on-nbcs-today-show/. The subject was originally slated to be cohost Meredith Viera, but her planned California trip was scrapped after the cancellation of the 2008 Golden Globe Awards.
10. Wojcicki, author interview, June 9, 2008.
11. Anne Wojcicki, "23andMe: Empowering consumer-enabled research," talk at UC Berkeley, June 26, 2008, http://www.youtube.com/watch?v=Q6KUf75bBEQ.
12. Stoffel is currently a professor at the Institute of Molecular Systems Biology in Zurich, Switzerland.
13. Linda Avey, "An American mother daughter legacy," *Huffington Post,* April 18, 2007, http://www.huffingtonpost.com/linda-avey-/an-american-mother-daught_b_46143.html.
14. Linda Avey, phone interview with author, April 11, 2007.
15. David A. Vise and Mark Malseed, *The Google Story* (New York: Delacorte Press, 2005).
16. Hugh Rienhoff, interview with author, Cambridge, Mass., November 14, 2007.
17. Andrew Pollack, "Company seeking donors of DNA for a 'gene trust,'" *New York Times,* August 1, 2000, http://www.nytimes.com/2000/08/01/business/company-seeking-donors-of-dna-for-a-gene-trust.html?scp=1&sq=%22DNA.com%22&st=cse.
18. Linda Avey, the Consumer Genetics Show, Boston, June 10, 2009.
19. Paul Berg was the third man to share the 1980 Nobel Prize in Chemistry, joining Fred Sanger and Walter Gilbert, who shared the other half for inventing methods for DNA sequencing.
20. Eric Lander, remarks at The Personal Genome: Consequences for Society conference, University of Washington, April 23, 2008, http://www.uwtv.org/programs/displayevent.aspx?rID=24551.

21. Technically there had been a few sporadic reports of association prior to 2005, including type 2 diabetes and a pair of genes for Crohn's disease.
22. The association between age-related macular degeneration and the gene for complement H was reported in three papers in the April 15, 2005, issue of *Science*.
23. Stephen P. Daiger, "Was the Human Genome Project worth the effort?" *Science* 308 (2005), 362.
24. Julius Gudmundsson et al., "Genome-wide association study identifies a second prostate cancer susceptibility variant at 8q24," *Nature Genetics* 39 (2007), 631–637.
25. The Wellcome Trust Case Control Consortium, "Genome-wide association study of 14,000 cases of seven common diseases and 3,000 shared controls," *Nature* 447 (2007), 661–678.
26. Brian Naughton, interview with author, Mountain View, Calif., June 9, 2008. Naughton helped write a selection of 23andMe white papers describing various methods used for ancestry and GWAS, http://www.23andme.com/for/scientists.
27. Timothy M. Frayling et al., "A common variant in the *FTO* gene is associated with body mass index and predisposes to childhood and adult obesity," *Science* 316 (2007), 889–894. See also Ellen Ruppel Shell, *The Hungry Gene* (New York: Atlantic Books, 2003).
28. The latest GWAS findings from more than 500 research papers recording some 2,600 SNPs (as of April 2010) are recorded online by the National Human Genome Research Institute at http://www.genome.gov/gwastudies.
29. Kevin Davies, "The Google genome connection," *Bio-IT World,* June 2007, www.bio-itworld.com/issues/2007/june/first-base.
30. Joanna Mountain, interview with author, Mountain View, Calif., June 9, 2008.
31. Christopher Anderson, *The Long Tail* (New York: Hyperion, 2008).
32. Avey, author phone interview.
33. Erin Cline, "Spit kit giveaway," The Spittoon, January 27, 2008, http://spittoon.23andme.com/2008/01/27/spit-kit-giveaway/.
34. Jay Flatley, Leerink Swann & Co., Emerging Products and Applications in Life Science Tools Conference, New York, August 7, 2007.
35. Goetz, "23andMe will decode."
36. Anne Wojcicki, "The power of We," The Spittoon, January 21, 2008, http://spittoon.23andme.com/2008/01/21/the-power-of-we/.
37. Linda Avey, "23andMe," keynote lecture, Bio-IT World Conference & Expo, Boston, April 30, 2008, http://www.bio-itworld.com/1sw/23andme.aspx.

38. Matt Crenson, "A beautiful ancestry painting," The Spittoon, June 13, 2008, http://spittoon.23andme.com/2008/06/13/a-beautiful-ancestry-painting/.
39. For example, 23andMe concluded that a 2007 *Science* paper reporting a gene association for a psychological trait called "avoidance of errors" warranted just one star. Likewise, a study on caffeine metabolism was viewed as preliminary rather than established research because of the small sample size.
40. Wojcicki, "The power of We."
41. Linda Avey and Anne Wojcicki, "Googling the Googlers' DNA," Google Tech Talk, May 7, 2008, http://www.youtube.com/watch?v=aeF-0y9HP9A.
42. Bloom syndrome is a recessive genetic disorder resulting in affected individuals having fragile chromosomes. The carrier frequency among people of Ashkenazi Jewish descent is about 1 in 100.
43. Wojcicki also underwent testing for the *BRCA1* breast cancer gene and found she was not a carrier. Again, using 23andMe, she learned that her genome was identical to both of her sisters in that region of chromosome 17, strongly suggesting that none is a *BRCA1* mutation carrier.
44. Wojcicki, "23andMe: Empowering consumer-enabled research."
45. Ibid.
46. Linda Avey, interview with author, Mountain View, Calif., June 9, 2008.
47. Allen Salkin, "When in doubt, spit it out," *New York Times*, September 12, 2008, http://www.nytimes.com/2008/09/14/fashion/14spit.html. Interestingly, the *Times* story appeared in the "Fashion & Style" section.
48. *Oprah*, November 14, 2008, http://www.oprah.com/oprahshow/Headline-Making-News/8.
49. Anita Hamilton, "The retail DNA test," *Time*, October 29, 2008, http://www.time.com/time/specials/packages/article/0.28804.1852747_1854493_1854113.00.html. 23andMe beat out inventions such as the Large Hadron Collider and the Chevy Volt.
50. Kara Swisher, "23andMe co-founder Linda Avey leaves personal genomics start-up to focus on Alzheimer's research," BoomTown, *Wall Street Journal* blog, September 4, 2009.
51. Anne Wojcicki, "Is your DNA your destiny?" TEDMED 2009, San Diego, October 27, 2009, http://www.tedmed.com/videos.

CHAPTER 3. EVERYBODY WANTS TO CHANGE THE WORLD

1. Kari Stefansson, phone interview with author, Reykjavik, June 27, 2008.
2. Stefansson's opinion of Craig Venter appears in *Cracking the Genome* (New York: Free Press, 2000), page 140.

3. Marcia Angell, *The Truth About the Drug Companies* (New York: Random House, 2004); Merrill Goozner, *The $800 Million Pill* (Berkeley: University of California Press, 2004).
4. At the close of 2008, deCODE hired a consulting firm to advise on a strategic reorganization but eventually filed for bankruptcy. A new entity bearing the deCODE name, founded on genetic research and diagnostics, emerged in 2010 under the leadership of former Genzyme executive Earl "Duke" Collier.
5. Kari Stefansson, deCODEme launch video, November 18, 2007, http://www.youtube.com/watch?v=Jug8-ie095E.
6. Gina Kolata, "New take on a prostate drug, and a new debate," *New York Times,* June 15, 2008, http://www.nytimes.com/2008/06/15/health/15prostate.html?_r=1&scp=3&sq=prostate%20cancer%20drugs%202008&st=cse.
7. *The Martha Stewart Show,* February 5, 2009, http://www.marthastewart.com/show/the-martha-stewart-show/actor-andy-garcia.
8. David B. Goldstein, *Jacob's Legacy: A Genetic View of Jewish History* (New Haven, CT: Yale University Press, 2008).
9. Nicholas Wade, "Genes show limited value in predicting diseases," *New York Times,* April 15, 2009, http://www.nytimes.com/2009/04/16/health/research/16gene.html?scp=1&sq=%22david%20b.%20goldstein%22&st=cse.
10. David B. Goldstein, "Common genetic variation and human traits," *New England Journal of Medicine* 360 (2009), 1696–1698; a rebuttal of sorts came in the same issue from Joel N. Hirschorn, "Genomewide association studies—illuminating biologic pathways," *New England Journal of Medicine* 360 (2009), 1699–1701.
11. Dennis Drayna, phone interview with author, August 12, 2009.
12. Jonathan C. Cohen et al., "Multiple rare variants in NPC111 associated with reduced sterol absorption and plasma low-density lipoprotein levels," *Proceedings of the National Academy of Sciences U.S.A.* 103 (2006), 1810–1815.
13. Malorye Branca, "Genome scan yields SIDS clue," *Bio-IT World,* August 2004, http://www.bio-itworld.com/archive/081804/sids.
14. Kevin Davies, "TGen's discovery pipeline in the desert," *Bio-IT World,* August 2005, http://www.bio-itworld.com/issues/2005/August-2005/bp-tgen.
15. Dietrich Stephan, phone interview with author, September 11, 2007.
16. Ron Winslow, "Is there a heart attack in your future?" *Wall Street Journal,* November 6, 2007, D1, http://online.wsj.com/article/SB119431099271083299-email.html.
17. Navigenics's communications director Brenna Sweeney kindly pro-

vided the Navitini recipe: 1 cup citrus vodka; ¼ cup fresh lemon juice; 3 tablespoons fresh pomegranate juice; ½ cup simple syrup (2 parts sugar, 1 part water); dash of rosewater; and garnish of orange peel.

18. Al Gore, speech at Navigenics launch party, New York, April 8, 2008.
19. Al Gore, James D. Watson Lecture, Washington, DC, January 20, 1998.
20. John Diamond, *C: Because Cowards Get Cancer Too* (New York: Vermilion, 1998).
21. David Agus, interview with author, Cedars-Sinai Medical Center, Los Angeles, February 22, 2009. Agus has since moved to the University of Southern California's Keck School of Medicine, where he is director of the Center for Applied Molecular Medicine.
22. Clifton Leaf, "Why we're losing the war on cancer," *Fortune,* March 22, 2004, http://money.cnn.com/magazines/fortune/fortune_archive/2004/03/22/365076/index.htm.
23. Dietrich Stephan, interview with author, Navigenics headquarters, Foster City, Calif., March 26, 2008.
24. The big three personal-genomics companies—Navigenics, 23andMe, and deCODEme—all launched their services before the passage of the Genetic Information and Nondiscrimination Act.
25. This is a prerequisite for most GWAS papers in the major journals.
26. Other criteria include the proper use of statistics for multiple hypothesis testing and appropriate matching of the control population, and it helps if there's some functional evidence surrounding the published variant.
27. Wellcome Trust Case Control Consortium, "Genome-wide association study of 14,000 cases of seven common diseases and 3,000 shared controls," *Nature* 447 (2007), 661–678.
28. Stephan, interview, March 26, 2008.
29. Navigenics customers must expressly request their raw SNP data, which can then be used in other programs such as Michael Cariaso's Promethease. Both 23andMe and deCODEme make the information available for download.
30. Winslow, "Is there a heart attack?"
31. Stephan, interview, March 26, 2008.
32. Agus, interview, February 22, 2009.

CHAPTER 4. DNA DREAMS

1. Mark D. Uehling, "Wanted: The $1000 genome," *Bio-IT World,* November 2002, http://www.bio-itworld.com/archive/111202/genome. The other speakers were Tony Smith (Solexa) and Michael Weiner (454 Life Sciences).

2. According to National Human Genome Research Institute administrator Jeffrey Schloss, his colleague Mark Guyer sent an e-mail on October 23, 2001, listing "$1,000 genome" as first on the list of proposed breakout session topics at the December 2001 NHGRI retreat.
3. Nick McCooke, interview with author, October 6, 2008.
4. Walter Gilbert, "A vision of the grail," in *The Code of Codes,* eds. Daniel J. Kevles and Leroy E. Hood (Cambridge: Harvard University Press, 1993), pp. 83–97.
5. Kevin Ulmer, interview with author, July 2, 2008. According to Ulmer, his original business plan for SEQ actually quoted the figure $1,200.
6. Eugene Chan, interview with author, September 8, 2007. Chan said, "I'd be sitting in Stu Schreiber's organic chemistry classes thinking, 'Dude, this is really easy!' It was kind of funny!"
7. Gary Stix, "Thinking big," *Scientific American,* June 2002, http://www.scientificamerican.com/article.cfm?id=thinking-big.
8. David Noonan, "Unraveling the genome," *Newsweek,* June 24, 2002, http://www.newsweek.com/id/64392.
9. Wil S. Hylton, "The gene jockey," *Esquire,* December 2002, http://www.esquire.com/features/best-n-brightest-2002/ESQ1202-DEC_CHANPENNINGER_rev_1.
10. Anon., "Eugene Chan, 29," *Technology Review,* October 2003, http://www.technologyreview.com/TR35/Profile.aspx?Cand=T&TRID=316.
11. Uehling, "Wanted."
12. Michael Shia, interview with author, Newton, Mass.: January 22, 2008.
13. "Jonathan M. Rothberg," retrieved January 1, 2008, from www.wikipedia.org. That entry also stated: "454 sequencing had a number of first (sic). It was the first new method to sequence a genome since Watson and Crick one (sic) the noble (sic) prize for sequencing in 1980 (sic)."
14. Jonathan Rothberg, interview with author, Sachem's Head, Conn., September 22, 2007.
15. Jonathan Rothberg, "Genomics 2.0," keynote lecture, Bio-IT World Conference & Expo, Boston, May 4, 2007.
16. Martin Leach, interview with author, Wellesley, Mass.: October 9, 2007.
17. Richard A. Shimkets et al., "Gene expression analysis by transcript profiling coupled to a gene database query," *Nature Biotechnology* 17 (1999), 798–803.
18. Laura Davis, phone interview with author, May 30, 2008.
19. Penelope Green, "The monoliths next door," *New York Times,* October 13, 2005, http://www.nytimes.com/2005/10/13/garden/13stone.html.
20. Michael Reis, "A grand vision becomes reality," *Stone World,* January 1, 2005, http://www.darrellpetit.com/articlesCircle.php.
21. Reis, "A grand vision."

22. Pal Nyren, "The history of pyrosequencing," *Methods in Molecular Biology* 73 (2006), 1–13.
23. Scott Helgesen, phone interview with author, February 14, 2008.
24. Andrew Pollack, "Company says it mapped genes of virus in one day." *New York Times,* August 22, 2003, http://www.nytimes.com/2003/08/22/business/company-says-it-mapped-genes-of-virus-in-one-day.html?scp=2&sq=virus%20rubin%20gibbs%202003&st=cse.
25. Koen Andries et al., "A diarylquinoline drug active on the ATP synthase of *Mycobacterium tuberculosis,*" *Science* 307 (2005), 223–227.
26. Marcel Margulies et al., "Genome sequencing in microfabricated high-density picolitre reactors," *Nature* 437 (2005), 376–380.
27. Jonathan Rothberg, e-mail to author, August 1, 2005.
28. Scott Helgesen, e-mail to author, February 17, 2008.
29. Matthias Krings et al., "Neandertal DNA sequences and the origin of modern humans," *Cell* 90 (1997) 19–30.
30. Nicholas Wade, "Scientists in Germany draft Neanderthal genome," *New York Times,* February 12, 2009, http://www.nytimes.com/2009/02/13/science/13neanderthal.html.
31. Richard E. Green et al., "A draft sequence of the Neandertal genome," *Science* 328 (2010), 710–722.
32. Stephan Schuster, Advances in Genome Biology and Technology conference, Marco Island, Fla., February 7, 2009. See also Erika Check Hayden, "Genome scan may save Tasmanian devils from cancer," *Nature News,* March 3, 2009, http://www.nature.com/news/2009/090303/full/news.2009.132.html.
33. Elizabeth Kolbert, "Stung," *The New Yorker,* August 6, 2007, http://www.newyorker.com/reporting/2007/08/06/070806fa_fact_kolbert.
34. Ibid.
35. Diana L. Cox-Foster et al., "A metagenomic survey of microbes in honey bee colony collapse disorder," *Science* 318 (2007), 283–287.
36. Gustavo Palacios et al., "A new arenavirus in a cluster of fatal transplant-associated diseases," *New England Journal of Medicine* 358 (2008), 991–998.
37. Robert Davis, "New genetic mapping technology IDs deadly virus," *USA Today,* February 7, 2008, http://www.usatoday.com/printedition/news/20080207/a_unravelingdisease07.art.htm.
38. Michael Egholm, interview with author, Guilford, Conn., February 15, 2008. Egholm moved to the Pall Corporation in 2010.
39. Rothberg, interview with author, Sachem's Head.

CHAPTER 5. THE BRITISH INVASION

1. Shankar Balasubramanian, interview with author, Cambridge, UK, November 21, 2008.
2. David R. Bentley et al., "Accurate whole human genome sequencing using reversible terminator chemistry," *Nature* 456 (2008), 53–59.
3. Another potential advantage of this so-called sequence-by-synthesis approach was that, unlike the 454 method, all four bases were added together, so that the DNA polymerase, as in nature, could select the appropriate base.
4. Klenerman and Balasubramanian filed the first patent, covering the broad concept of preparing a single-molecule array for sequencing.
5. John Milton, interview with author, Boston, November 3, 2008.
6. Harold Swerdlow, phone interview with author, October 16, 2008.
7. Helen Briggs, "Your genetic code on a disk," BBC News online, September 23, 2002, http://news.bbc.co.uk/2/hi/science/nature/2276095.stm. A year later, McCooke gave a talk at a genome ethics meeting in Heidelberg and predicted the possible arrival of "the 1000 Euro genome" by the end of 2010.
8. Ido Braslavsky et al., "Sequence information can be obtained from single DNA molecules," *Proceedings of the National Academy of Sciences U.S.A.* 100 (2003).
9. There was one small wrinkle: patent issues required reinventing the amplification method, changing from PCR to an isothermal bridge amplification process. See http://www.illumina.com/documents/products/datasheets/datasheet_cbot.pdf.
10. Clive Brown, phone interview with author, October 3, 2008.
11. John West, phone interview with author, February 11, 2009.
12. Robert M. Frederickson, "Just bead it," *Bio-IT World,* February 18, 2004, http://www.bio-itworld.com/archive/021804/equipped.html.
13. Kevin Davies, "Solexa completes full virus genome sequence," *Bio-IT World,* April 2005, http://www.bio-itworld.com/issues/2005/april/sci-solexa/.
14. John West, interview with author, Hilton Head, South Carolina, October 21, 2005. West's back-of-the-envelope calculation over breakfast went like this: The human genome is three billion bases, and resequencing an individual requires coverage of at least fifteen times, or 45 billion bases. Two days per run means 90 days for a genome on a single machine. The flow cells cost $1,000–$3,000, so 45 flow cells would cost at least $45,000. Depreciation for a $400,000 instrument over five years (three months per genome) adds another $20,000. Other personnel and overhead pushed the estimate to around $100,000.

15. Brendan Borrell, "First Asian genome announced," *Nature News,* October 23, 2007, http://www.nature.com/news/2007/071012/full/news.2007.161.html.
16. Bentley later added a further 48 gigabases in another thirteen runs.
17. Kevin Davies, "Now and future sequencing stories," *Bio-IT World,* June 10, 2008, http://www.bio-itworld.com/BioIT_Article.aspx?id=76560&LangType=1033.
18. The Illumina GA II had crisper images, higher resolution, better illumination, better read quality, and redesigned flow cells. A 4 megapixel CCD camera imaged the same flow cell area in 100 (rather than 300) tiles, decreasing the cycle time to 90 minutes (from 130) and the run time from just over 3 days to just over 2.
19. Jay Flatley, phone interview with author, November 3, 2009.
20. For perspective, the global repository of DNA sequence data, GenBank, had just one-fifth that amount, or 200 gigabases.
21. Kevin Davies, "Illumina wins this round on points." *Bio-IT World,* January/February 2009, http://www.bio-itworld.com/issues/2009/ian-feb/illumina-wins-first-round.html. At the end of 2008, the Sanger Institute elected to return its ABI machines and invest in eleven more Illumina instruments; the pipelines and staff were already optimized for that platform.
22. Jay Flatley, remarks at GET (Genomes, Environments, Traits) Conference, Cambridge, Mass., April 29, 2010.
23. Carolyn Abraham, "X marks the spotlight for elusive benefactor," *Toronto Globe & Mail,* October 13, 2006, http://www.theglobeandmail.com/news/technology/science/article850255.ece.
24. Stewart Blusson, phone interview with author, December 11, 2007.
25. Vernon Frolick, *Fire Into Ice* (Vancouver: Raincoast Books, 2002).
26. Jonathan Rothberg, interview with author, Sachem's Head, Conn., September 22, 2007.
27. Hawking's quote at the X Prize Web site, http://www.xprize.org.
28. The list of X Prize for Genomics entrants is at www.xprize.org. There were eight entrants as of May 2010: Roche/454, VisiGen Biotechnologies (owned by Life Technologies), Foundation for Applied Molecular Evolution, Reveo, base4 Innovation, Personal Genome X-team, ZS Genetics, and Cracker (Taiwan).
29. Staff, "Helicos teams with ISB," *Red Herring,* November 15, 2006, http://www.redherring.com/Home/19797.
30. Andrew Pollack, "The race to read genomes on a shoestring, relatively speaking," *New York Times,* February 9, 2008, http://www.nytimes.com/2008/02/09/business/09genome.html?pagewanted=1&_r=1.
31. Gordon Sanghera, interview with author, San Diego, April 23, 2008.

CHAPTER 6. SERVICE CALL

1. Kevin Davies, "The next-gen data deluge," *Bio-IT World,* April 2008. http://www.bio-itworld.com/issues/2008/april/cover-story-dna-data-deluge.html.
2. Kevin Davies, *Cracking the Genome* (New York: Free Press, 2000).
3. Davies, "Data deluge." The two-base encoding method used a pool of short, single-stranded DNA cassettes just eight bases in length. The method searches pairs of nucleotides (16 possible combinations of 4 x 4 bases) with four fluorescent tags. Blanchard assigned one color to four dinucleotide combinations. So long as the identity of the first base is known the method can deduce the sequence by sliding along the DNA one base at a time. The method can distinguish a real mutation, which will affect the color of two tags, versus a random sequencing error.
4. Michael Rhodes, Exploring Next-Generation Sequencing, San Diego, March 21, 2007.
5. Kevin Davies, "Girl power," *Bio-IT World,* April 2008, http://www.bio-itworld.com/issues/2008/april/next-gen-sequencing.html.
6. Kevin Judd McKernan et al., "Sequence and structural variation in a human genome uncovered by short-read, massively parallel ligation sequencing using two base encoding," *Genome Research* 19 (2009), 1527–1541. ABI's first next-generation human genome sample was the same African HapMap volunteer that Illumina published six months earlier.
7. Kevin McKernan, memorandum to the House of Lords Select Committee on Science and Technology, November 5, 2008, http://www.parliament.uk/documents/upload/101stGMAppliedBiosystems.pdf. McKernan noted that some senior Sanger Institute staff such as Harold Swerdlow used to work for Solexa, while Illumina's chief scientist, David Bentley, had moved the other way, creating at least the appearance of a conflict of interest.
8. The Broad Institute's Stacy Gabriel is using Illumina for high-throughput cancer genome sequencing work, part of the 1,000 Genomes Project. At CHI's Next-Generation Sequencing conference in Providence, Rhode Island, September 23, 2009, Gabriel said that Illumina instruments required much less DNA as the starting material than SOLiD.
9. According to David Dooling of The Genome Center at Washington University, St. Louis, it was impractical to maintain the resources to run two high-throughput platforms. Another factor was that the SOLiD platform required larger quantities of DNA compared to Illumina, which was often a factor in cancer research. See Kevin Davies, "David Dooling: Gangbusters at The Genome Center," *Bio-IT World,* 2009, http://www.bio-itworld.com/BioIT_Article.aspx?id=94067.

10. Stanley Lapidus, interview with author, Cambridge, Mass., June 28, 2007.
11. Igor Braslavsky et al., "Sequence information can be obtained from single DNA molecules," *Proceedings of the National Academy of Sciences U.S.A.* 100 (2003), 3960–3964.
12. Stanley Lapidus, UBS Global Life Sciences Conference, New York, September 25, 2007.
13. Lapidus interview.
14. Keith J. Winstein, "DNA decoding maps mainstream future," *Wall Street Journal,* October 4, 2007, http://online.wsj.com/article/SB119144984347648190.html.
15. Bill Efcatvitch, remarks of Advances in Genome Biology and Technology, Marco Island, Fla., February 8, 2008.
16. Stephen Quake, The Genetics of Common Diseases, Broad Institute, September 9, 2008.
17. Bill Efcavitch, Advances in Genome Biology and Technology, Marco Island, Fla., February 7, 2009.
18. Stephen Quake, "Genome mania," *New York Times,* March 3, 2009, http://opinionator.blogs.nytimes.com/2009/03/03/guest-column-genome-mania/.
19. Dmitry Pushkarev et al., "Single-molecule sequencing of an individual human genome," *Nature Biotechnology* 27 (2009), 847–850, http://thebigone.stanford.edu/quake/publications/pushkarev_nbt.pdf.
20. Bruce Martin, interview with author, Mountain View, Calif., February 23, 2009. See Kevin Davies, "Complete compute: An interview with Bruce Martin," *Bio-IT World,* September–October 2009, http://www.bio-itworld.com/2009/09/23/martin.html.
21. Andrew Pollack, "Dawn of low-price mapping could broaden DNA uses," *New York Times,* October 6, 2008, http://www.nytimes.com/2008/10/06/business/06gene.html?_r=1.
22. Kevin Davies, "Complete Genomics targets 'the first $1000 genome,' " *Bio-IT World,* November–December 2008, http://www.bio-itworld.com/issues/2008/nov-dec/complete-genomics.html. The DNA constructs took several days to prepare, although the final step of making the nanoballs could be done in thirty minutes.
23. Clifford Reid, phone interview with author, October 1, 2008.
24. James Hudson, phone interview with author, April 6, 2009.
25. Kevin Ulmer, interview with author, Cohasset, Mass., July 2, 2008.
26. Laura DeFrancesco, "Profile: Rade Drmanac," *Nature Biotechnology* 26 (2008), 1100.
27. Rade Drmanac, interview with author, Marco Island, Fla., February 8, 2009.

28. Rade Drmanac et al., "DNA sequence determination by hybridization: A strategy for efficient large-scale sequencing," *Science* 260 (1993), 1649–1652.
29. HySeq merged with Variagenics and is now called Nuvelo.
30. Drmanac created a holding company, Callida Genomics, with about a dozen scientists and began working in earnest on a new sequencing strategy.
31. Drmanac, interview with author, Mountain View, Calif., February 23, 2009.
32. Each DNA nanoball contained thousands of copies of the same circle of template DNA, including seventy bases of the sequence to be read. A complete description of the Complete Genomics sequencing strategy can be found at its Web site, http://www.completegenomics.com.
33. Drmanac interview, Mountain View.
34. Lee Hood, phone interview with author, October 3, 2008.
35. Clifford Reid, interview with author, Mountain View, Calif., February 23, 2009.
36. Clifford Reid, Advances in Genome Biology and Technology, Marco Island, Fla., February 7, 2009.
37. Dramanc interview, Marco Island.
38. Clifford Reid, keynote lecture, Bio-IT World Conference and Expo, Boston, April 29, 2009.
39. Ibid.
40. Ibid.
41. Clifford Reid, phone interview with author, August 24, 2009.
42. Reid, keynote lecture.
43. Rade Drmanac, interview with author, Honolulu, Hawaii, October 22, 2009. Complete's commercial instrument achieved a 100-fold increase in throughput by improving the density of the nanoballs and improved optics. The product upgrades allowed Complete to skip an entire generation of planned instruments.
44. The family with Miller syndrome—the Madsens of Utah—was analyzed by two groups. The Complete Genomics paper: Jared C. Roach et al., "Analysis of genetic inheritance in a family quartet by whole-genome sequencing," *Science* 328 (2010), 636–639. Another analysis was reported by researchers at the University of Washington: Sarah B. Ng et al., "Exome sequencing identifies the cause of a Mendelian disorder," *Nature Genetics* 42 (2010), 30–35.
45. Radoje Drmanac et al., "Human genome sequencing using unchained base reads on self-assembling DNA nanoarrays," *Science* 327 (2010), 78–81.
46. Clifford Reid, interview with author, Marco Island, February 27, 2010.

CHAPTER 7. MY GENOME AND ME

1. Elissa Levin, interview with author, Redwood City, Calif., March 27, 2008. Navigenics provided me with a complementary test.
2. Dietrich Stephan, interview with author, Redwood City, Calif., March 27, 2008.
3. Kevin Davies, "Cellf examination," *Bio-IT World,* May 2005, http://www.bio-itworld.com/issues/2005/april/first-base.aspx.
4. Anne Underwood and Jerry Adler, "Diet and genes." *Newsweek,* January 17, 2005, http://www.newsweek.com/id/48235.
5. Carina Dennis, "Rugby team converts to give gene tests a try," *Nature* 434 (2005), 260.
6. Kevin Davies, "Genes, geography and history," *Bio-IT World,* June 2005, http://www.bio-itworld.com/issues/2005/June/document.2005-06-10.8675816706.
7. Timothy M. Frayling et al., "A common variant in the *FTO* gene is associated with body mass index and predisposes to childhood and adult obesity," *Science* 316 (2007), 889–894.
8. Indeed they did change over time. In August 2009, my risk had crept up to 32 percent.
9. Don Winslow, "Is there a heart attack in your future?" *Wall Street Journal,* November 6, 2007, http://online.wsj.com/article/SB119431099271083299-email.htmlhttp://online.wsj.com/article/SB119431099271083299-email.html.
10. deCODE generously provided me with a complementary deCODEme test.
11. John N. Feder et al., "A novel MHC class I-like gene is mutated in patients with hereditary haemochromatosis," *Nature Genetics* 13 (1996), 399–408.
12. Kari Stefansson, phone interview with author, June 28, 2008.
13. David Ewing Duncan, *The Experimental Man* (Hoboken, N.J.: Wiley, 2009); Francis Collins, *The Language of Life* (New York: HarperCollins, 2010); see also Pauline C. Ng et al., "An agenda for personalized medicine," *Nature* 461 (2009), 724–726.
14. Jon McClellan and Mary-Claire King, "Genetic heterogeneity in human disease," *Cell* 141 (2010), 210–217.
15. James Plante, phone interview with author, July 14, 2009.
16. The Hempel Web site is at www.addiandcassi.com.
17. Chris Hempel's quest to find a cure for her daughters' disease is described in several reports, including Amy Dockser Marcus, "A mom brokers treatment for her twins' fatal illness," *Wall Street Journal,* April 3, 2009, http://online.wsj.com/article/SB123871183055784317.html.

18. David Becker, phone interview with author, July 14, 2009.
19. Maurissa Bornstein, Pathway Genomics, e-mail to author, November 18, 2009.
20. Michael Cariaso, phone interview with author, September 30, 2008.
21. As of March 2010, SNPedia held more than 10,000 SNPs of proven or suspected medical relevance. In principle, dbSNP, the full catalog of human SNPs, could hold about 20 million SNPs.
22. Michael Corino, "User" Lilly Mendel, http://www.snpedia.com/index.php/User:Lilly_Mendel.
23. Linda Avey, *The Life and Times of Lilly Mendel,* November 2009. Avey elected to reveal her identify because some of her friends, using 23andMe's genome-sharing feature, had compared Avey's public SNP data with those of "Lilly Mendel" and found a 100 percent match.
24. Michael Cariaso, interview with author.
25. James D. Watson, "First lessons from my personal genome," Double Helix Medals dinner, New York City, November 6, 2008, http://www.dana.org/news/features/detail.aspx?id=18970.
26. Sarina M. Rodrigues et al., "Oxytocin receptor genetic variation relates to empathy and stress reactivity in humans," *Proceedings of the National Academy of Sciences USA* 106 (2009), 21437–21441.
27. My full Promethease report can be found here at SNPedia: http://www.snpedia.com/index.php/User:Natgenex.
28. Michael Christman, remarks at the Consumer Genetics Show, Boston, Mass., June 4, 2010.
29. Kevin Davies, "Pinpointing a perfect pitch gene," *Bio-IT World,* August 13, 2009, http://www.bio-itworld.com/news/08/13/09/pinpointing-perfect-pitch-gene.html.

CHAPTER 8. CONSUMER REPORTS

1. "The Killer in You," *60 Minutes* (Australia), May 29, 2009. http://video.ninemsn.com.au/video.aspx?mkt=en-AU&brand=ninemsn&vid=5b8570bb-e24f-45e3-81f9-fc7c0b7d87bc#::5b8570bb-e24f-45e3-81f9-fc7c0b7d87bc.
2. "A conversation about personalized medicine," *Charlie Rose Show,* PBS, June 19, 2009, http://www.charlierose.com/view/interview/10399.
3. Thomas Goetz, "23andMe will decode your DNA for $1,000: Welcome to the age of genomics," *Wired,* November 2007, http://www.wired.com/medtech/genetics/magazine/15-12/ff_genomics.
4. Mark Henderson, "Handle with care: Genetic tests are risky, and I've got the proof," *Times,* March 1, 2008, http://www.timesonline.co.uk/tol/news/science/article3463550.ece.

5. Boonsri Dickinson, "Inside out: A DNA diary," *Discover,* September 2008, http://discovermagazine.com/2008/sep/20-how-much-can-you-learn-from-a-home-dna-test.
6. A handful of deCODEme customer profiles can be found at http://www.decodeme.com/customer-stories. In one case, a twenty-seven-year-old Canadian student, Anna Peterson, motivated by her mother's breast cancer diagnosis, learned she had a moderately increased risk for the disease, a very low risk of age-related macular degeneration despite a family history, and a surprisingly high (46 percent) lifetime risk for basal cell carcinoma. She later learned that her grandfather had died of skin cancer. She is now armed with knowledge to be vigilant.
7. Amy Harmon, "My genome, myself: Seeking clues in DNA," *New York Times,* November 17, 2007, http://www.nytimes.com/2007/11/17/us/17dna.html.
8. "Cracking the code," *Nightline,* ABC News, February 14, 2008, http://abcnews.go.com/Video/playerIndex?id=4288707.
9. Cassandra Jardine, "DNA testing: A dip in the gene pool reveals what I'm made of," *Daily Telegraph,* June 2, 2008, http://www.telegraph.co.uk/health/dietandfitness/3354986/DNA-testing-A-dip-in-the-gene-pool-reveals-what-Im-made-of.html.
10. Esther Dyson, "The altruism of genetic testing," *Big Think,* May 2009, http://bigthink.com/ideas/15536.
11. *Charlie Rose Show,* "Personalized medicine."
12. "Headline making news," *Oprah,* November 14, 2008, http://www.oprah.com/oprahshow/Live-with-Rob-Lowe-Melissa-Etheridge-and-More/9.
13. David Ewing Duncan, *Experimental Man* (Hoboken, N.J.: Wiley, 2009).
14. Francis Collins, keynote lecture at the Consumer Genetics Show, Boston, June 8, 2009.
15. Francis Collins, *The Language of Life* (New York: HarperCollins, 2010).
16. Geoffrey Smigelsky, "dnaSNIPs: A comparison study into consumer genetics services," retrieved November 18, 2009, from http://dnasnips.com.
17. Pauline C. Ng et al., "An agenda for personalized medicine," *Nature* 461 (2009), 724–726.
18. Linda Avey, *The Life and Times of Lilly Mendel,* November 4, 2009, http://lillymendel.blogspot.com/2009/11/recent-craig-venter-et-al.html.
19. Peter Aldhous, "Gene predictions tell an ever-changing story," *New Scientist,* July 29, 2009, http://www.newscientist.com/article/mg20327194.600-gene-predictions-tell-an-everchanging-story.html.
20. The *Times*'s Mark Henderson reported the same phenomenon: when

deCODEme added new SNPs for cardiovascular disease, Henderson's risk of heart attack rose above 60 percent.

21. "Terry Moran's moment of truth," *ABC Nightline,* May 5, 2009, http://www.truveo.com/Terry-Moran%E2%80%99s-Moment-of-Truth/id/767901956.
22. Anna Gosline, "What's in your DNA?" *Los Angeles Times,* April 14, 2008, http://articles.latimes.com/2008/apr/14/health/he-genome14.
23. Ricki Lewis, "A virtual pharma organization," *Bio-IT World,* June 2008, http://www.bio-itworld.com/issues/2008/june/allen-roses-interview.html.
24. "Terry Moran's moment of truth."
25. Jeffrey Drazen, "The Road to Personal Genomics" roundtable, Bio-IT World Conference and Expo, April 30, 2008. http://www.bio-itworld.com/lsw/personalgenomics.aspx.
26. Robert Green, phone interview with author, October 22, 2008.
27. Robert C. Green et al., "Disclosure of APOE genotype for risk of Alzheimer's disease," *New England Journal of Medicine* 361 (2009), 245–254.
28. Eric S. Lander et al., "Initial sequencing and analysis of the human genome," *Nature* 409 (2001), 860–921.
29. Eric Lander, speaking at The Personal Genome: Consequences for Society conference, University of Washington, April 23, 2008, http://www.uwtv.org/programs/displayevent.aspx?rID=24551.
30. Carey Goldberg, "Some doubt genome's value as health tool," *Boston Globe,* April 21, 2008, http://www.boston.com/news/science/articles/2008/04/21/some_doubt_genomes_value_as_health_tool/.
31. Trisomy 18 is a severe congenital disorder caused by inheriting three copies of chromosome 18. (Trisomy 21 is commonly known as Down syndrome.)
32. George Church's page at the Personal Genomes Project Web site: http://www.personalgenomes.org/public/1.html.
33. George Church personal history and interests: http://arep.med.harvard.edu/gmc/pers.html.
34. The estimate of $10 per base pair comes from J. G. Sutcliffe and G. Church, "The cleavage site of the restriction endonuclease Ava II," *Nucleic Acids Research* 5 (1978), 2313–2319. In 2009, Complete Genomics charged customers $20,000 for a human genome (30 billion bases), or about $0.000001 per base pair. And as Church noted, that was the retail price, not the cost.
35. George M. Church, "The personal genome project," *Molecular Systems Biology* 1 (2005), 0030, http://www.nature.com/msb/journal/v1/n1/full/msb4100040.html.
36. Editorial, "Capitalizing on the genome," *Nature Genetics* 13 (1994),

1–4, http://www.nature.com/ng/wilma/v13n1.867861436.html. Church worked with Genome Therapeutics to sequence the genome of the bacterium *Helicobacter pylori,* which causes peptic ulcers. The work was not published, and it was largely forgotten when Craig Venter published the first microbial genome the following year.

37. *Polonies* were pools of amplified DNA molecules ("PCR colonies") embedded in a gel that provided an excellent template for DNA sequencing.
38. According to PubMed, the term *personal genomics* was first published in Gary Stix's *Scientific American* article about U.S. Genomics's Eugene Chan: "Innovations: Thinking big," June 2002.
39. George M. Church, "Genomes for all," *Scientific American,* January 2006.
40. Rosie Mestal, "Anonymous sperm donor traced on Internet," *New Scientist,* November 3, 2005, http://www.newscientist.com/article.ns?id=mg18825244.200. The boy sent a cheek swab to a genealogy service called FamilyTreeDNA.com. Nine months later, he was contacted independently by two men with Y-chromosome profiles almost identical to his own—and the same surname. Because the boy happened to know his biological father's date and place of birth, he acquired a list of everyone born in that place and day using Omnitrace.com. Only one man had the matching surname. Ten days later, he had made contact with his biological father.
41. Church, "Genomes." Church thought the PGP would fare better as a nonprofit entity rather than a company (such as the ill-fated DNA Sciences) and solicited support from all sources, including insurance companies, which, he thought a tad optimistically, would recognize "a potential sea change." No word on whether that has come to pass.
42. Ibid.
43. The PGP 10 are PGP-1, George Church; PGP-2, John Halamka (chief information officer, Harvard Medical School); PGP-3, Esther Dyson; PGP-4, Misha Angrist (Duke University); PGP-5, Kirk Maxey (CEO, Cayman Chemical); PGP-6, Steven Pinker (professor of psychology, Harvard University); PGP-7, Keith Batchelder (CEO, Genomic Healthcare); PGP-8, Stanley Lapidus (founder and chairman, Helicos Biosciences); PGP-9, Rosalyn Gill (former CEO, Sciona); and PGP-10, James Sherley (Boston Biomedical Research Institute). www.personalgenomes.org/pgp10.html.
44. Esther Dyson, "Full disclosure," *Wall Street Journal,* July 25, 2007.
45. Stanley Lapidus, interview with author, Cambridge, Mass., January 29, 2008.
46. John Halamka, "The Road to Personal Genomics" panel, Bio-IT World Expo, April 30, 2008, http://www.bio-itworld.com/lsw/personalgenomics.aspx.

47. Steven Pinker, "My genome, my self," *New York Times Magazine*, January 7, 2009, http://www.nytimes.com/2009/01/11/magazine/11Genome-t.html.
48. Trait-o-matic screens several databases, including Online Mendelian Inheritance in Man (OMIM), the Human Genome Mutation Database (HGMD), and SNPedia, which is free to use for noncommercial research purposes. OMIM and HGMD focus on mutations associated with genetic disease; by contrast, most of the SNPs in SNPedia were highlighted by association studies, not direct mutation analysis.
49. George Church, phone interview with author, September 8, 2009. Some putative mutations highlighted by Trait-o-matic are likely to be sequencing errors, particularly, says Church, "if you find what looks like an incredibly deleterious mutation and the person is perfectly healthy and walking around."
50. Ibid. Neither Church nor the cardiologist identified the PGP individual by name, but, given the central tenet of the PGP, it was not a secret. "We're trying to treat it as if it doesn't matter, until it does," said Church. "Generally what matters is the relationship between trait and gene, not the relationship between trait and name." The Trait-o-matic results for the PGP 10 were posted online in October 2009 at the PGP Web site: snp.harvard.edu.
51. Christine Seidman, e-mail to author, August 7, 2009.
52. Pinker, "My genome, my self."
53. Green, interview.

CHAPTER 9. CEASE AND DESIST

1. Linda Avey, presentation at the Secretary's Advisory Committee on Genetics, Health and Society, Washington, DC, February 12, 2008, http://oba.od.nih.gov/SACGHS/sacghs_past_meeting_documents.html.
2. Ann Willey, presentation at CHI's Beyond Genome conference, San Francisco, June 11, 2008.
3. The thirteen companies targeted by CDPH were DeCode Genetics, Knome, Navigenics, 23andMe, CGC Genetics, DNATraits, Gene Essence, HairDX, New Hope Medical, Salugen, Sciona, Smart Genetics, and Suracell. The letters were sent out rather sloppily. The letter to 23andMe was addressed to "Linda Avery," and the letter to "DeCodeMe Genetics" assumed that the CEO was a woman: "Ms Kari Stefansson." "It doesn't take particularly close inspection to determine what sex I am," Stefansson said. "I'm not particularly feminine."
4. Lea Brooks, CDPH, e-mail to author, June 17, 2008.

5. David Agus, interview with author, Beverly Hills, Calif., February 22, 2009.
6. Ryan Phelan, "DNA Direct in full compliance with California regulations," DNA Direct Talk, June 2006, http://talk.dnadirect.com/2008/06.
7. Thomas Goetz, "Attention, California Health Dept.: My DNA is my data," *Wired Science,* June 17, 2008, http://www.wired.com/wired science/2008/06/attention-calif/.
8. Kevin Davies, "Personal genomics companies rush to comply with California regulations," *Bio-IT World,* June 18, 2008, http://www.bio-itworld .com/personal-genomics-companies-rush-to-comply.html.
9. An earlier draft of the statement declared the tests were "ordered and reviewed" by a physician, but the draft was later modified. Consumers could order the Navigenics and other personal genomics tests without consulting a physician.
10. Kari Stefansson, phone interview with author, June 25, 2008.
11. W. Bradley Tulley, letter to Karen L. Nickel, June 23, 2008.
12. Andrew Pollack, "California licenses two companies to offer gene services," *New York Times,* August 19, 2008, http://www.nytimes.com/2008/ 08/20/business/20gene.html.
13. Anne Willey, presentation at the Secretary's Advisory Committee on Genetics, Health, and Society, Washington, DC, March 12, 2009, http://oba .od.nih.gov/SACGHS/sacghs_past_meeting_documents.html.
14. Willey said another sticking point was that consumer-genomics firms would not be allowed if there was a financial relationship with the ordering physician.
15. Potential 23andMe customers ordering "New York-Bound Spit Kits" received this notice: "Upon receipt of your Spit Kit, you or the Spit Kit recipient will be required to affirm under penalty of law that the sample for the Spit Kit has not been collected in or mailed from the state of New York."
16. Linda Avey, interview with author, Boston, November 19, 2009.
17. Anne Wojcicki, interview with author, January 8, 2010.
18. "Navigenics receives State of New York clinical laboratory permit," *Navigator,* January 11, 2010, http://blog.navigenics.com/articles/comments/ navigenics_receives_state_of_new_york_clinical_laboratory_permit/.
19. Muin Khoury, interview with author, August 28, 2009.
20. David J. Hunter, Muin J. Khoury, and Jeffrey M. Drazen, "Letting the genome out of the bottle: Will we get our wish?" *New England Journal of Medicine* 358 (2008), 105–107.
21. Muin Khoury, "Interview with Muin Khoury on personal genomics services being offered directly to consumers," *New England Journal of Medicine,* http://content.nejm.org/cgi/content/full/358/2/105/DC1.

22. Privacy fears were largely allayed by the passage of the Genetic Information Nondiscrimination Act (GINA) in 2008.
23. Hunter et al., "Letting the genome out of the bottle."
24. Robert Green, phone interview with author, October 22, 2008.
25. Peter Aldhous, "My 'non-human' DNA: A cautionary tale," *New Scientist,* August 26, 2009, http://www.newscientist.com/article/dn17683-my-nonhuman-dna-a-cautionary-tale.html.
26. Avey, interview, November 19, 2009.
27. Jeffrey M. Drazen, "The road to personal genomics," roundtable at the Bio-IT World Conference and Exposition, Boston, April 30, 2008, http://www.bio-itworld.com/lsw/personalgenomics.aspx.
28. Sharon Terry helped researchers identify the first gene mutated in PXE. O. Le Saux et al., "Mutations in a gene encoding an ABC transporter cause pseudoxanthoma elasticum," *Nature Genetics* 25 (2000), 223–227.
29. Philip R. Reilly, presentation at the Consumer Genetics Show, Boston, June 7, 2009.
30. Sharon R. Terry, testimony at the House Ways and Means Committee, Subcommittee on Health, March 14, 2007.
31. For GINA, the House voted 414 to 1. The Senate voted 95 to 0. The legislation almost passed in 2005, when the U.S. Senate voted in favor 95 to 0, but it was blocked by Senator Tom Coburn (R-Oklahoma). Currently two states have stricter laws than GINA: Massachusetts and South Dakota.
32. Associated Press, "Conn. woman alleges genetic discrimination at work," *Boston Herald,* April 28, 2010, http://www.bostonherald.com/news/national/northeast/view/20100428conn_woman_alleges_genetic_discrimination_at_work/srvc=home&position=recent.
33. Robert C. Green and George J. Annas, "The genetic privacy of presidential candidates," *New England Journal of Medicine* 359 (2008), 2192–2193, http://content.nejm.org/cgi/content/full/359/21/2192.
34. William R. Brody, "A brave new insurance," *Wall Street Journal,* December 20, 2002.
35. Stephen J. Dubner, "Genetics entrepreneur Anne Wojcicki answers your questions," *New York Times,* August 12, 2009, http://freakonomics.blogs.nytimes.com/2009/08/12/genetics-entrepreneur-anne-wojcicki-answers-your-questions/?pagemode=print.
36. Kathy Hudson et al., "ASHG statement on direct-to-consumer genetic testing in the United States," *American Journal of Human Genetics* (September 2007), 635–637.
37. American College of Medical Genetics, "What the public needs to know about direct-to-consumer genetic tests," April 24, 2008, http://bit.ly/acmg08.

38. Harry Ostrer, "The personal genome: What we know now," *Big Think,* July 29, 2009, http://bigthink.com/paulhoffman/full-discussion-of-the-personal-genome.
39. Larry Thompson, presentation to the Secretary's Advisory Committee on Genetics, Health and Society, Washington, DC, March 12, 2009, http://oba.od.nih.gov/SACGHS/sacghs_past_meeting_documents.html.
40. Juliet Macur, "Born to run? Little ones get test for sports gene," *New York Times,* November 29, 2008, http://www.nytimes.com/2008/11/30/sports/30genetics.html.
41. Daniel MacArthur, "The ACTN3 sports gene test: What can it really tell you?" *Genetic Future,* November 30, 2008, http://scienceblogs.com/geneticfuture/2008/11/the_actn3_sports_gene_test_wha.php.
42. "Laws of attraction," *The Early Show,* CBS Television, October 23, 2008, http://www.cbsnews.com/video/watch/?id=4541364n.
43. My Gene Profile's eight categories are IQ, emotional quotient, athletic ability, character, health, environmental sensitivity, creativity, and addiction susceptibility. www.mygeneprofile.com.
44. "At home genetic tests: A healthy dose of skepticism may be the best prescription," *FTC Facts of Consumers,* July 2006, http://www.ftc.gov/bcp/edu/pubs/consumer/health/hea02.pdf.
45. John Sulston, remarks at "The Personalized Genome: Do I Want to Know?" Gairdner Foundation Fiftieth Anniversary, University of Toronto, October 30, 2009, http://mediacast.ic.utoronto.ca/20091030-GFS/index.htm#.
46. Mary K. Engle, letter to Rosalynn Gill, August 14, 2009, http://www.ftc.gov/os/closings/090814scionaclosingletter.pdf.
47. Dan Vorhaus, "What five FDA letters mean for the future of DTC genetic testing," *Genomics Law Report,* June 11, 2010, http://bit.ly/FDAlet5.

CHAPTER 10. ANOTHER WEEK, ANOTHER GENOME

1. Other pronunciation variations are "K-Nome" and "K-Nam-me."
2. Jorge Conde, interview with author, Cambridge, Mass., January 29, 2008.
3. Just as the genome is the sum total of all our genetic material, *exome* is the term given to the sum total of all the genes in the genome, or about 1 to 2 percent of the total DNA. Genes are themselves divided into exons (the coding regions) and spacers called introns. Most of the mutations we care about fall in genes, altering the structure or function of the proteins they encode. Thus, as a practical matter, sequencing the exome is a high-value, more affordable alternative to sequencing the entire genome.

(A similar approach is to sequence the *transcriptome,* or all of the expressed genes.) The problem is that no one can be sure what will be missed in all the DNA that is not sequenced.

4. Conde, interview.
5. Elizabeth Pennisi, "Breakthrough of the year: Human genetic variation," *Science* 318 (2007), 1842–1843, http://www.sciencemag.org/cgi/content/full/318/5858/1842.
6. Amy Harmon, "Gene map may become a luxury item," *New York Times,* March 4, 2008, http://www.nytimes.com/2008/03/04/health/04iht-04geno.10677657.html.
7. Ibid.
8. Knome preferred using clients' blood rather than saliva to ensure it obtained sufficient DNA and avoid the potential embarrassment of going back to request another sample.
9. Richard Powers, "The book of me," *GQ* (October 2008), http://www.gq.com/news-politics/big-issues/200810/richard-powers-genome-sequence.
10. Richard P. Ebstein et al., "Dopamine D4 receptor (D4DR) exon III polymorphism associated with the human personality trait of novelty seeking," *Nature Genetics* 12 (1996), 78–80; Jonathan Benjamin et al., "Population and familial association between the D4 dopamine receptor gene and measures of novelty seeking," *Nature Genetics* 12 (1996), 81–84.
11. Michael Snyder et al., "Personal genome sequencing: Current approaches and challenges," *Genes and Development* 24 (2010), 423–431, http://www.genesdev.cshlp.org/content/24/5/423.full.
12. Julie Steenhuysen, "Glenn Close has genes mapped," Reuters, March 11, 2010, http://www.reuters.com/article/idUSTRE62A5P620100311.
13. Gert-Jan van Ommen, interview with author, June 5, 2009. The Solexa instrument was originally earmarked for 1 billion bases per run, but the Leiden group "on a lucky day" could do more than 2 gigabases. "Gee," van Ommen thought, "that's almost a whole human genome."
14. Although the male Y chromosome is a shriveled excuse of a human chromosome, carrying only a few dozen functional genes, there is still reason to study it, not least because mutations on the Y chromosome have been associated with infertility. See Steve Jones, *Y: The Descent of Man* (Boston: Houghton Mifflin, 2003).
15. Without access to Illumina's paired-end sequencing at this time and very short read lengths, Kriek's genome was very difficult to assemble, with thousands of gaps.
16. This wasn't the first time *Fokke & Sukke* had dabbled in medical genetics. In an earlier cartoon, the pair posed as genetic counselors. While Fokke peered down a microscope, Sukke addressed a nervous couple: "Con-

gratulations, he's going to be a healthy Dutch lad . . . But unfortunately, in the 2022 World Cup final, his shot hits the goalpost!"

17. Marjolein Kriek, phone interview with author, June 5, 2009. As of June 2010, the Kriek genome had not been published.
18. Illumina press briefing at the American Society of Human Genetics conference, Philadelphia, November 11, 2008. The November 5, 2008, issue of *Nature* carried three articles on the first African genome, the first Asian genome, and the first cancer genome, all featuring Solexa sequencing on the Illumina platform. David R. Bentley et al., "Accurate whole human genome sequencing using reversible terminator chemistry," *Nature* 456 (2008), 53–59; Jun Wang et al., "The diploid sequence of an Asian individual," *Nature* 456 (2008), 60–65; and Timothy J. Ley et al., "DNA sequencing of a cytogenetically normal acute myeloid leukaemia genome," *Nature* 456 (2008), 66–72.
19. In general, the Chinese are more reticent to identify subjects because individuals do not truly have complete ownership of their DNA: they share parts of the same sequence with their children and relatives. Identifying the "owner" of the DNA would be considered extremely unethical.
20. BGI is sequencing 100 Asian genomes for the 1000 Genome Project, including the genome of a Chinese billionaire who donated $10 million to BGI, and has embarked on its own 1000 Genome Project for the Asian population.
21. Saeed Hussain, interview with author, January 8, 2009.
22. According to Hussain, the five populations for the Arabian genome project are (1) original Arabs from Yemen; (2) people in contact with original Arabs who came from the north (Syria and Jordan, for example), including the Prophet Muhammad; (3) Egyptians; (4) North Africans (Algeria, Morocco, Tunisia); and (5) African Arabians. Saudi Arabia and the Gulf region would contribute at least half of the volunteers.
23. Ruiquiang Li et al., "The sequence and de novo assembly of the giant panda genome," *Nature* 463 (2009), 311–317. Significantly, the BGI team showed it was possible to assemble a mammalian genome from scratch even with short DNA reads.
24. Jay Flatley, remarks at the GET Conference, Cambridge, Mass., April 27, 2010.
25. Elaine R. Mardis et al., "Recurring mutations found by sequencing an acute myeloid leukemia genome," *New England Journal of Medicine* 361 (2009), 1058–1066.
26. By contrast, the first AML genome required ninety-eight runs, with shorter read lengths, achieving only 91 percent coverage.

27. Dan Koboldt, "Second cancer genome in *New England Journal,*" MassGenomics, August 6, 2009, http://www.massgenomics.org/2009/08/second-cancer-genome-in-new-england-journal.html.
28. The reason Kahn referred to four genomes rather than three is that the St. Louis group effectively sequenced a pair of genomes: the tumor and the matched normal tissue control.
29. Korean Reference Genome Project, www.koreagenome.org.
30. Sung-Min Ahn, e-mail to author, June 7, 2009.
31. Sung-Min Ahn et al., "The first Korean genome sequence and analysis: Full genome sequencing for a socio-ethnic group," *Genome Research* 19 (2009), 1622–1629, http://genome.cshlp.org/content/19/9/1622.full.
32. Jong-Il Kim et al., "A highly annotated whole-genome sequence of a Korean individual," *Nature* 460 (2009), 1011–1015.
33. Kevin McKernan, e-mail to author, August 7, 2009. Responding to other criticism that he had not submitted to *Nature,* he said *Genome Research* had offered him more space in a special issue devoted to personal genomics to describe the analysis and the technology.
34. Kevin McKernan, comment to "ABI SOLiD joins the WGS party," MassGenomics, July 2, 2009, http://www.massgenomics.org/2009/07/abi-solid-joins-the-wgs-party.html.
35. Marfan syndrome is a dominantly inherited disorder, so the only way Rienhoff's daughter could have acquired the disease was if it resulted from a new mutation.
36. Online Mendelian Inheritance in Man, http://www.ncbi.nlm.nih.gov/sites/entrez?db=omim.
37. Brendan Maher, "Personal genomics: His daughter's DNA," *Nature* 449 (2007), 773–776, http://www.nature.com/news/2007/071017/full/449773a.html.
38. Bart L. Loeys et al., "A syndrome of altered cardiovascular, craniofacial, neurocognitive and skeletal development caused by mutations in TGFBR1 or TGFBR2," *Nature Genetics* 37 (2005), 275–281.
39. Cattle and mice with mutated myostatin genes develop grotesque bodybuilder physiques, as if they'd overdosed on steroids.
40. Anon., "Genetics, medicine and insurance," *Economist,* August 23, 2007.
41. Brendan I. Koerner, "DIY DNA: One father's attempt to hack his daughter's genetic code," *WIRED* (February 2009), http://www.wired.com/medtech/genetics/magazine/17-02/ff_diygenetics?currentPage=all.
42. Ibid.
43. Maher, "Personal genomics."
44. Ibid.
45. Koerner, "DIY DNA."

46. Hugh Rienhoff, "My daughter's DNA," Google Tech talk, November 29, 1997, http://www.youtube.com/watch?v=4WoaQhjWmRU.
47. Dimitry Pushkarev, Norma F. Neff, and Stephen R. Quake, "Single-molecule sequencing of an individual human genome," *Nature Biotechnology* 27 (2009), 847–850.
48. Stephen R. Quake, "Genome mania," *New York Times,* March 3, 2009, http://opinionator.blogs.nytimes.com/2009/03/03/guest-column-genome-mania/.
49. Kevin Ulmer, e-mail to author, August 8, 2009.
50. As an investigator with (and technically an employee of) the Howard Hughes Medical Institute, Quake was forbidden to collaborate with companies ("one of their Ten Commandments" was how he put it). See Kevin Davies, "The single life: Stephen Quake Q&A," *Bio-IT World,* August 10, 2009, http://www.bio-itworld.com/news/08/10/09/stephen-quake-interview.html.
51. Quake found more than 2.8 million SNPs, of which almost 25 percent were novel, as well as 752 copy number variants (CNVs) spanning 16 megabases.
52. Euan A. Ashley et al., "Clinical assessment incorporating a personal genome," *Lancet* 375 (2010), 1525–1535. Although clearly a preliminary study, the paper provided a proof of principle that clinically relevant data could be extracted from an individual's whole genome sequence.
53. David Dooling, "Another rich white guy's genome," *PolITIGenomics,* August 11, 2009, http://www.politigenomics.com/2009/08/another-rich-white-guy-sequences-own-genome.html.
54. Kevin Davies, "What can Brown do for Oxford Nanopore?" *Bio-IT World,* September 2009, http://www.bio-itworld.com/NGS-Brown.html.
55. Jay Flatley, interview with author, the Consumer Genetics Show, Boston, June 10, 2009.
56. Kevin Davies, "The $48,000 man: Illuminating Hermann Hauser's genome," *Bio-IT World,* September 9, 2009, http://www.bio-itworld.com/BioIT_Content.aspx?id=94250&terms=2007.
57. Patients could get information from a new Web site, www.everygenome.com, and participate in social networking for the education and exchange of information for those who have had their genomes sequenced. Illumina is also establishing a network of partners to offer a variety of services involved in data analysis and counseling, as well as to the physicians themselves.
58. Kevin Davies, "Illumina drops personal genome sequencing price to below $20,000," *Bio-IT World,* June 3, 2010, http://bit.ly/ILMNigs.
59. James R. Lupski et al., "DNA duplication associated with Charcot-

Marie-Tooth disease type 1A," *Cell* 66 (1991), 219–232. CMT was first described by three physicians in the 1860s, including Jean-Martin Charcot, the father of modern neurology and a mentor to medical icons including Georges Gilles de la Tourette, Joseph Babinksi, and Sigmund Freud.

60. The filtering step removes SNPs already cataloged in the database (and thus considered too common to be the root cause of a rare genetic disorder), those found in HapMap individuals, and SNPs in noncoding (junk) DNA.
61. Jim Lupski, interview with author, Honolulu, October 22, 2009.
62. Kirsten Stewart, "Unraveling one family's genome," *Salt Lake City Tribune,* March 10, 2010, http://www.sltrib.com/utah/ci_14648608.
63. Murim Choi et al., "Genetic diagnosis by whole exome capture and massively parallel DNA sequencing," *Proceedings of the National Academy of Sciences USA* 106 (2009), 19096–19101. Lifton used the exome approach rather than sequencing the patient's entire genome.
64. Lupski said *Nature* had invited him to submit the paper, but in the meantime it had published a proof-of-principle study from Jay Shendure's group (University of Washington) demonstrating the ability to identify a mutation for a rare disorder by sequencing all the protein-coding genes. The paper was finally published as James R. Lupski et al., "Whole-genome sequencing in a patient with Charcot-Marie-Tooth neuropathy," *New England Journal of Medicine,* March 10, 2010.
65. Stephan C. Schuster et al., "Complete Khoisan and Bantu genomes from southern Africa," *Nature* 463 (2010), 943–947.
66. Radoje Drmanac, interview with author, Marco Island, Fla., February 27, 2010.
67. Clifford Reid, interview with author, Marco Island, Fla., February 27, 2010.
68. George Church, interview, *Charlie Rose Show,* PBS, June 19, 2009, http://www.charlierose.com/view/interview/10399.
69. George Church, phone interview with author, September 8, 2009.
70. The best major histocompatibility complex marker for narcolepsy is HLA-DQB1*0602, which is carried by more than 90 percent of narcoleptic patients. But this HLA variant is neither necessary nor sufficient for the disorder.
71. Snyder et al., "Personal genome sequencing."
72. K. G. Skryabin et al., "Combining two technologies for full genome sequencing of human," *Acta Naturae* 3 (2009), http://actanaturae.ru/article.aspx?id=173.

CHAPTER 11. THE 15-MINUTE GENOME

1. There are too many third-generation sequencing companies to do justice to. Almost without exception, these technologies are exciting and imaginative, and some may well become mainstream commercial successes or find applications tangential to personal genome sequencing. See Michael L. Metzker, "Sequencing technologies—the next gneration," *Nature Reviews Genetics* 11 (January 2010), 31–46.
2. William Glover, "Base pair discrimination via transmission electron microscope," presentation at CHI's Next-Generation Sequencing Case Studies and Applications, San Diego, April 24, 2008. Glover had hoped to have a prototype ready in 2009 but that deadline came and went.
3. Halcyon called its method IMPRNT (Individual Molecule Placement Rapid Nano Transfer) and claimed it could leapfrog over second-generation approaches. Initial testing was done using a $2 million electron microscope on loan from the U.S. Department of Energy.
4. Jerzy Olejnik, "The Pinpoint sequencer," CHI's Next-Generation Sequencing, San Diego, April 24, 2009. The ultimate goal of 20 billion bases per day sounded impressive, but Illumina and Life Technologies reached that throughput nine months later.
5. John Markoff, "I.B.M. joins pursuit of $1,000 genome sequencing," *New York Times,* October 5, 2009, http://www.nytimes.com/2009/10/06/science/06dna.html?_r=1&scp=1&sq=IBM%20nanopore&st=cse.
6. Andrew Pollack, "The race to read genomes on a shoestring, relatively speaking," *New York Times,* February 9, 2008, http://www.nytimes.com/2008/02/09/business/09genome.html.
7. Steve Turner, interview with author at the Advances in Genome Biology and Technology conference, Marco Island, Fla., February 9, 2008.
8. Hugh Martin became CEO in 2004 following the series A financing.
9. Suzanne Barlyn, "Name that firm," *Wall Street Journal,* March 17, 2008, http://online.wsj.com/article/SB120526710337728101.html/.
10. Michael J. Levene et al., "Zero-mode waveguides for single-molecule analysis at high concentrations," *Science* 299 (2003), 682–686.
11. Actually about one-third of the wells, according to the Poisson distribution. Turner said that PacBio would eventually provide a kit that allows enzymes to fall into every well.
12. By piggybacking the tag on the end of the nucleotide, the fluorophore sits in the enzyme's active site during the catalytic step but then is clipped away, spinning off into the void like a space-walking astronaut who severs his life support. What gets left behind is a perfectly natural growing strand of DNA.
13. Mardis later joined PacBio's scientific advisory board.

14. John Eid et al., "Real-time DNA sequencing from single polymerase molecules," *Science* 323 (2009), 133–138.
15. Michael V. Copeland, "Gene machine," *Fortune,* September 29, 2008, http://money.cnn.com/2008/09/26/technology/pacific_biosciences.fortune/index.htm.
16. Kevin Davies, "PacBio nets Eric Schadt as chief scientific officer," *Bio-IT World,* May 28, 2009, http://www.bio-itworld.com/news/2009/05/28/schadt-to-pacbio.html.
17. Julia Karow, "PacBio reveals commercial specs; initial focus on long reads, short runs, low experiment cost," *In Sequence,* November 17, 2009.
18. Matthew Herper, "The new, fast gene machine," *Forbes,* October 5, 2009, http://www.forbes.com/forbes/2009/1005/revolutionaries-science-genomics-gene-machine.html.
19. The SMRT cells are lined into strips of eight (dubbed "eight-packs"); twelve of these strips can be loaded at a time in a ninety-six-well format. Each SMRT cell is individually sealed, so an instrument run could involve just a single SMRT cell, returning to the other cells in the eight-pack strip later.
20. The strobe mode overcomes photophysical damage to the polymerase inflicted by excited fluorophores that occasionally "go into a bad state," as Martin puts it.
21. Kevin Davies, "Splash down: Pacific Biosciences unveils third-generation sequencing machine," *Bio-IT World,* February 26, 2010, http://www.bio-itworld.com/2010/02/26/pacbio.html.
22. The company was formerly known as Oxford NanoLabs.
23. The Dyson Perrins building dated back to World War I and was established by the heir to the Lea and Perrins Worcestershire sauce fortune.
24. Hagan Bayley, interview with author, Oxford University, November 20, 2008.
25. Langzhou Song et al., "Structure of staphylococcal alpha-hemolysin, a heptameric transmembrane pore," *Science* 274 (1996), 1859–1866.
26. Hagan Bayley, "Building doors into cells," *Scientific American* (September 1997), 62–67.
27. John J. Kasianowicz et al., "Characterization of individual polynucleotide molecules using a membrane channel," *Proceedings of the National Academy of Sciences USA* 93 (1996), 13770–13773.
28. Paul Durman, "Norwood takes up university challenge," *Sunday Times,* April 9, 2006, http://business.timesonline.co.uk/tol/business/industry_sectors/support_services/article703372.ece.
29. Yann Astier, Orit Braha, and Hagan Bayley, "Toward single molecule DNA sequencing: Direct identification of ribonucleoside and deoxyri-

bonucleoside 5'-monophosphates by using an engineered protein nanopore equipped with a molecular adapter," *Journal of the American Chemical Society* 128 (2006), 1705–1710.

30. Gordon Sanghera, interview with author, San Diego, April 22, 2008.
31. Exonucleases chew nucleotides off either end of a DNA strand. This is the same enzyme that Kevin Ulmer proposed to use more than a decade earlier as the centerpiece of his single-molecule sequencing company SEQ.
32. Jennifer Griffiths, "The realm of the nanopore," *Analytical Chemistry* 80 (2009), 23–27.
33. James Clarke et al., "Continuous base identification for single-molecule nanopore DNA sequencing," *Nature Nanotechnology* 4 (2009), 265–270.
34. Kevin Davies, "What can Brown do for Oxford Nanopore?" *Bio-IT World,* September 22, 2009, http://www.bio-itworld.com/NGS-Brown.html.
35. John Milton, interview with author, Boston, October 2, 2008.
36. Kevin Davies, "It's 'Watson meets Moore' as Ion Torrent introduces semiconductor sequencing," *Bio-IT World,* March 1, 2010, http://www.bio-itworld.com/news/03/01/10/ion-torrent-semiconductor-sequencing.html.
37. If two identical bases are incorporated, the chip detects a doubling of the pH change. That linear response is accurate for runs of at least six identical bases, and Rothberg says it is important to get to accurately call eight consecutive polymeric bases. Beyond that, the repeat is probably not biologically relevant, he asserted.
38. Jonathan Rothberg, interview with author, Marco Island, Fla., February 26, 2010.
39. The purines (A and G) were depicted with closed symbols, whereas pyrimidines (C and T) were open.

CHAPTER 12. PERSONALIZED RESPONSE

1. BiDil is a combination of two generic drugs: isosorbide dinitrate and hydralazine (I/H). The discovery of nitric oxide (NO) as a potent biochemical signaling molecule won the Nobel Prize in Physiology or Medicine in 1998 and led to the development of Viagra and other erectile dysfunction drugs, which prevent NO breakdown.
2. Jay N. Cohn et al., "A comparison of enalapril with hydralazine-isosorbide dinitrate in the treatment of chronic congestive heart failure," *New England Journal of Medicine* 325 (1991), 303–310.
3. Jay Cohn, interview with author, May 9, 2006. ACE inhibitors did not appear to serve African Americans as well as whites, possibly due to re-

duced activity of the NO system, which might also explain the increased incidence of hypertension in blacks.

4. Peter Carson et al., "Racial differences in response to therapy for heart failure: Analysis of the vasodilator-heart failure trials," *Journal of Cardiac Failure* 5 (1999), 178–187.
5. Pamela Sankar and Jonathan Kahn, "BiDil: Race medicine or race marketing?" *Health Affairs,* October 11, 2005, http://content.healthaffairs.org/cgi/content/full/hlthaff.w5.455/DC1.
6. Anne L. Taylor et al., "Combination of isosorbide dinitrate and hydralazine in blacks with heart failure," *New England Journal of Medicine* 351 (2004), 2049–2057. See also Steve Stiles, "Dr. Jay N. Cohn and A-HeFT: Persistence rewarded," TheHeart.org, November 5, 2004, http://www.theheart.org/article/359499.do.
7. Jonathan Kahn, "Putting BiDil in context: Law, commerce, and racial drugs," presentation at The Business of Race and Science, MIT, April 7, 2006.
8. Keith Ferdinand, "BiDil and race: Practical considerations and future implications," presentation at The Business of Race and Science, MIT, April 7, 2006.
9. Susan Reverby, "BiDil as Tuskegee's child: What does it mean?" presentation at The Business of Race and Science, MIT, April 7, 2006.
10. Jonathan Kahn, "How a drug becomes ethnic: Law, commerce, and the production of racial categories in medicine," *Yale Journal of Health Policy, Law and Ethics* 4 (2004), 1–46.
11. Steven Nissen, phone interview with author, June 3, 2006.
12. Kevin Davies, "The race prescription card," *Bio-IT World,* January 2005, http://www.bio-itworld.com/archive/012105/firstbase.html.
13. Stephan C. Schuster et al., "Complete Khoisan and Bantu genomes from southern Africa," *Nature* 463 (2010), 943–947.
14. *Genetics and Race,* science writers workshop at the American Society of Human Genetics meeting, Toronto, October 27, 2004. See "Genetics for the human race," *Nature Genetics* 36 (2004) S1–S60.
15. Henry Louis Gates, *Faces of America,* PBS, 2010, http://www.pbs.org/wnet/facesofamerica/.
16. Cohn, interview.
17. David B. Goldstein and Huntington F. Willard, "Race, the genome," *Boston Globe,* January 17, 2005, http://boston.com/news/globe/editorial_opinion/oped/articles/2005/01/17/race_the_genome.
18. Kevin Davies, "FDA approves first 'ethnic medicine,' " *Bio-IT World,* June 2005, http://www.bio-itworld.com/newsitem/2005/06-05-06-24-05-news-bidil.
19. Anna Helgadottir et al., "A variant of the gene encoding leukotriene A4

hydrolase confers ethnicity-specific risk of myocardial infarction," *Nature Genetics* 38 (2006), 68–74. The HapK variants were found in 20 percent of African-American patients compared to just 6 percent of healthy controls, a 250 percent increase in risk.

20. One theory is that in older European populations, gene variants may have had time to counteract HapK, which may have risen to prominence by conferring protection against infectious disease hundreds of years ago.
21. Marcia Angell, *The Truth About Drug Companies* (New York: Random House, 2004).
22. Eric P. Hoffman, "Skipping towards personalized molecular medicine," *New England Journal of Medicine* 357 (2007), 2719–2722.
23. John Carey, "The maker of a miracle drug," *Business Week,* December 24, 2008, http://www.businessweek.com/innovate/content/dec2008/id20081224_690553.htm?chan=innovation_auto+design_voices+of+innovation. Gleevec has earned Novartis more than $3 billion annually, proving that blockbuster drugs are possible even when the patient population is narrowed using some form of diagnostic or molecular test.
24. Raphael Lehrer, e-mail to author, May 11, 2010.
25. Kevin Davies, "Jay Tenenbaum urges collaboration to treat the long tail of disease," *Bio-IT World,* February 26, 2009, http://www.bio-itworld.com/2009/02/26/tenenbaum-mmtc-keynote.html.
26. Anthony Fejes, "Sequencing cancer genomes and transcriptomes: From new pathology to cancer treatment," Fejes.ca, February 7, 2009, http://www.fejes.ca/2009_02_01_archive.html. The drug was Sutent.
27. Many of these SNPs were featured in the SNpedia top ten list of SNPs for 2009: http:/www.snpedia.com.
28. Leroy Hood, "A doctor's vision of the future of medicine," *Newsweek,* July 13, 2009, http://www.newsweek.com/id/204227.
29. Daniel S. Levine, "Gazing into the future," *Journal of Life Sciences* (Fall 2009), 18–25, http://www.burrillreport.com/content/pastissues/TJOLS_Fall2009.pdf#page=20.
30. Hood, "A doctor's vision of the future of medicine."
31. Luke Nosek, e-mail, September 24, 2009, http://www.commonparadox.com/2009/09/paypal-co-founder-and-founders-fund-partner-joins-dna-sequencing-firm-halcyon-molecular/.
32. Stanley Lapidus, interview with the author, Cambridge, Mass., January 29, 2008.
33. Epigenetic changes of the genome likely play a major role in controlling gene activity in response to developmental or environmental changes.

CHAPTER 13. THE REST OF US

1. *Nature Genetics* Question of the Year, 2007: "What would you do if it became possible to sequence the equivalent of a full human genome for only $1000?" http://www.nature.com/ng/qoty/index.html.
2. The NHGRI had an annual budget for DNA sequencing of $120 million just before Collins resigned in 2008.
3. Kevin Davies, "New CMS method pinpoints positive selection signatures in human populations," *Bio-IT World,* January 7, 2010, http://www.bio-itworld.com/BioIT_Article.aspx?id=96268&terms=2007. See also Sharon R. Grossman et al., "A composite of multiple signals distinguishes causal variants in regions of positive selection," *Science* 327 (2010), 883–886.
4. Mark Henderson, "Genetic map of tumours reveals thousands of mutations," *Times,* December 17, 2009, http://www.timesonline.co.uk/tol/news/uk/health/article6959691.ece. Another lung cancer genome study by Genentech and Complete Genomics estimated one mutation for every three cigarettes smoked.
5. Jeffrey Gulcher and Kari Stefansson, "Genetic risk information for common diseases may indeed be already useful for prevention and early detection," *European Journal of Clinical Investigation* 40 (2010), 56–63.
6. Lee M. Silver, *Remaking Eden* (New York: HarperCollins, 2007).
7. Alan H. Handyside et al., "Pregnancies from biopsied human preimplantation embryos sexed by Y-specific DNA amplification," *Nature* 344 (1990), 768–770.
8. Editorial, "Life after SuperBabe," *Nature* 454 (2008), 253.
9. Peter Little, *Genetic Destinies* (New York: Oxford University Press, 2001).
10. Helen Pearson, "Making babies: The next 30 years," *Nature* 454 (2008), 260–262.
11. The list of disorders currently amenable for PGD at Genesis Genetics is at http://www.genesisgenetics.org/testing.html.
12. Mark Hughes, interview with author, January 8, 2010.
13. Kate Devlin, "Birth of first British baby genetically screened for breast cancer," *Daily Telegraph,* January 9, 2009, http://www.telegraph.co.uk/health/healthnews/4208538/Birth-of-first-British-baby-genetically-screened-for-breast-cancer.html.
14. Lorraine Fisher, "I've designed my baby," *Daily Mail,* November 14, 2006, http://www.dailymail.co.uk/health/article-416297/Ive-designed-baby.html.
15. Sarah-Kate Templeton, "Doctors screen embryos to avoid babies with

squint," *Sunday Times,* May 6, 2007, http://www.timesonline.co.uk/tol/life_and_style/health/article1752351.ece.

16. Carina Dennis, "Genetics: Deaf by design," *Nature* 431 (2004), 894–896.
17. In twenty years, Hughes said he has never been asked to help a couple have a child with a disease, and he doesn't believe that other U.S. clinics have been asked either.
18. Stan Lapidus, interview with author, Cambridge, Mass., January 29, 2008.
19. Jim Kelly, quoted in March of Dimes press release, March 27, 2009, http://www.huntershope.org/site/DocServer/032709_Jim_Kelly_Urges_AZ_Legislature.pdf?docID=1564.
20. 3MCC stands for 3-methylcrotonyl-CoA carboxylase.
21. www.savebabies.org.
22. Kevin Davies, "A legacy for and beyond Batten disease," *Bio-IT World* (March–April 2010), http://www.bio-itworld.com/BioIT_Article.aspx?id=97662.
23. John N. Feder et al., "A novel MHC class I-like gene is mutated in patients with hereditary haemochromatosis," *Nature Genetics* 13 (1996), 399–408. "Haemochromatosis . . . definite maybe!" *Nature Genetics* 13 (1996), 375–376, http://www.nature.com/ng/journal/v13/n4/pdf/ng0896-375.pdf. Given the element of doubt surrounding the gene discovery, the title of the editorial borrowed the phrase a conflicted Rebecca exhaled after her first kiss with Sam on *Cheers.*
24. Dennis Drayna, phone interview with author, August 12, 2009.
25. Anne Wojcicki, interview with author, Mountain View, Calif., June 9, 2008.
26. Eric Lander, remarks at The Personal Genome: Consequences for Society, University of Washington, Seattle, April 23, 2008, http://www.uwtv.org/programs/displayevent.aspx?rID=24551.
27. John McPherson, "Next-generation gap," *Nature Methods* 6 (2009), S2–S5.
28. Dean Hamer, *The God Gene* (New York: Doubleday, 2004). Hamer made headlines in 1995 with a paper in *Science* mapping a putative "gay gene" on the X chromosome. Hamer's suggestion that a variant in the *VMAT2* gene could explain how faith is hard-wired into our genetic makeup has not been published in a peer-reviewed study.
29. Changsoo Kang et al., "Mutations in the lysosomal enzyme-targeting pathway and persistent stuttering," *New England Journal of Medicine* 362 (2010), 677–685.
30. Steve Connor, "Glaxo chief: Our drugs do not work on most patients," *Independent,* December 8, 2003, http://www.independent.co.uk/news/

science/glaxo-chief-our-drugs-do-not-work-on-most-patients-575942.html.

31. Sydney Brenner, remarks at The Personalized Genome: Do I Want to Know? Gairdner Foundation 50th Anniversary, Toronto, October 30, 2009, http://mediacast.ic.utoronto.ca/20091030-GFS-P4/index.htm.
32. Charles Sabine, remarks at The Personalized Genome: Do I Want to Know? Gairdner Foundation 50th Anniversary, Toronto, October 30, 2009, http://mediacast.ic.utoronto.ca/20091030-GFS-P4/index.htm.
33. Bill Gates, remarks at The Personal Genome: Consequences for Society, University of Washington, April 23, 2008. Gates said, "After we cure the top 20 infectious diseases, then I'll get my genome sequenced. . . . Absolutely!" http://www.uwtv.org/programs/displayevent.aspx?rID=24551.
34. John Sulston, remarks at The Personalized Genome: Do I Want to Know? Gairdner Foundation 50th Anniversary, Toronto, October 30, 2009, http://mediacast.ic.utoronto.ca/20091030-GFS-P4/index.htm.
35. Anne Wojcicki, phone interview with author, January 5, 2010.
36. GET (Genomes, Environments, Traits) Conference, Cambridge, Mass., April 27, 2010.
37. Jong Bhak, e-mail to author, April 30, 2010. Bhak was a coauthor on the second Korean genome paper published in 2009.

Acknowledgments

The $1,000 Genome is a story of and a tribute to the remarkable efforts of scores of talented scientists and pioneers, many of whom generously and repeatedly gave me their time and shared their insights and knowledge. I particularly want to thank David Agus, Linda Avey, Shankar Balasubramanian, David Becker, David Bentley, Jong Bhak, Mark Boguski, John Boyce, Clive Brown, Han Cao, Mike Cariaso, Eugene Chan, George Church, Jorge Conde, Gina Costa, Dennis Drayna, Rade Drmanac, Bill Efcavitch, Michael Egholm, Richard Fisler, Richard Gibbs, Rosalynn Gill, Stan Gloss, Philip Goelet, Jim Golden, Jeff Gulcher, Eric Green, Robert Green, John Halamka, Susan Hardin, Tim Harkins, Tim Harris, Hermann Hauser, Scott Helgesen, Mark Hughes, Hywel Jones, Scott Kahn, Muin Khoury, Bruce Korf, Marjolein Kriek, Stanley Lapidus, Martin Leach, Fred Ledley, Samuel Levy, Steve Lombardi, Jim Lupski, Nick McCooke, Kevin McKernan, Christopher McLeod, John Milton, Joanna Mountain, Ken Offit, Ryan Phelan, Richard Powers, Egor Prokhortchouk, Stephen Quake, Philip Reilly, Jacques Retief, Hugh Rienhoff, Jonathan Rothberg, Gordon Sanghera, Christine Seidman, Jay Shendure, Mike Shia, Tony Smith, Kari Stefansson, Dietrich Stephan, Mark Stevenson, Kristen Stoops, Harold Swerdlow, Saeed Al Turki, Stephen Turner, Kevin Ulmer, Gert-Jan van Ommen, Jun Wang, Anne and John West, David Wheeler, Stefan White, Spike Willcocks, Anne Wojcicki, Xiaodi Wu, and Tian Xu.

Thanks also to Maurissa Bornstein, Wes Conard, Kaustuva Das, Chris d'Eon, Ed Farmer, Jane Kramer, Nicole Litchfield, Lisa Osborne, Jon Schmid, Fintan Steele, Brenna Sweeney, Martha Trela, Jennifer Turcotte, Patty Zamora, and especially Zoe McDougall, who made my job possible and fun.

I also have to thank my fellow journalists—and the bloggers who often put us to shame: Peter Aldhous, Mark Henderson, Julia Karow, Daniel MacArthur, and Dan Vorhaus among others.

I am in the unusual and fairly fortunate position to have not one but two genomic counselors, and must praise Elissa Levin and Emily Enns for their trailblazing efforts. I also thank deCODE, Navigenics, Pathway Genomics, and SeqWright for providing complementary genome scans.

This book would not have materialized without the extraordinary patience and support of Emily Loose, my gifted and tireless editor at Free Press, not to mention Gill Kent, Edith Lewis, and Beverly H. Miller for their copyediting heroics. Thanks also to my wonderful agent, Jen Gates at Zachary Shuster Harmsworth. I am indebted to Phillips Kuhl, president of Cambridge Healthtech Institute (CHI), for supporting this project without hesitation, along with many other dedicated colleagues at CHI and *Bio-IT World,* especially Allison Proffitt, Mary Ann Brown, Mark Gabrenya, and Alan El Faye. And for their continued encouragement, I'm really grateful to Laurie Goodman, Bette Phimister, John Russell, Oona Snoeyenbos-West, and Amanda Wren.

Above all, thanks to my family for their extraordinary patience and forbearance during the past three years. This book exists thanks to you!

Index

About the Author

Kevin Davies, PhD, is the author of *Cracking the Genome.* He is currently Editor-in-Chief of *Bio-IT World,* a monthly magazine covering enabling technology in the life sciences. He was the founding editor of *Nature Genetics,* the world's leading genetics journal, which he headed for its first five years. He has also written for the *Times* (London), *Boston Globe, New England Journal of Medicine, New Scientist,* and *Prospect.* Davies holds an MA in biochemistry from the University of Oxford and a PhD in molecular genetics from the University of London. He lives in Lexington, Massachusetts, with his wife and two teenage children.